I0896950

3D Printing and Digital Twins in Smart Agriculture

P. William
Sanjivani University, India

Sharmila A.
Raj Kumar Goel Institute of Technology, India

A. K. Haghi
University of Coimbra, Portugal

IGI Global
Scientific Publishing
Publishing Tomorrow's Research Today

Vice President of Editorial Melissa Wagner
Director of Acquisitions Mikaela Felty
Director of Book Development Jocelynn Hessler
Production Manager Mike Brehm
Cover Design Jose Rosado

Published in the United States of America by
IGI Global Scientific Publishing
701 East Chocolate Avenue
Hershey, PA, 17033, USA
Tel: 717-533-8845 | Fax: 717-533-7115
Website: https://www.igi-global.com E-mail: cust@igi-global.com

Copyright © 2026 by IGI Global Scientific Publishing. All rights reserved. No part of this publication may be reproduced, stored or distributed in any form or by any means, electronic or mechanical, including photocopying, without written permission from the publisher. Use of this publication to train generative artificial intelligence (AI) technologies is expressly prohibited. The publisher reserves all rights to license its use for generative AI training and machine learning model development.

Product or company names used in this set are for identification purposes only. Inclusion of the names of the products or companies does not indicate a claim of ownership by IGI Global Scientific Publishing of the trademark or registered trademark.

Library of Congress Cataloging-in-Publication Data

Names: William, P. editor | Sharmila A., 1985- editor | Haghi, A. K. editor

Title: 3D printing and digital twins in smart agriculture / edited by P. William, Sharmila A., A.K. Haghi.
Description: Hershey, PA : IGI Global Scientific Publishing, 2026. | Includes bibliographical references and index. | Summary: "This book explores how 3D printing and digital twin technologies are revolutionizing smart agriculture. It highlights innovations that enhance precision farming, resource efficiency, and sustainability, offering insights into how these tools can transform agricultural practices, improve productivity, and address global food security challenges"-- Provided by publisher.
Identifiers: LCCN 2025041860 (print) | LCCN 2025041861 (ebook) | ISBN 9798337370774 hardcover | ISBN 9798337370781 paperback | ISBN 9798337370798 ebook
Subjects: LCSH: Agricultural innovations | Three-dimensional printing | Digital twins (Computer simulation)
Classification: LCC S494.5.I5 A117 2026 (print) | LCC S494.5.I5 (ebook)
LC record available at https://lccn.loc.gov/2025041860
LC ebook record available at https://lccn.loc.gov/2025041861
British Cataloguing in Publication Data
A Cataloguing in Publication record for this book is available from the British Library.

All work contributed to this book is new, previously-unpublished material.
The views expressed in this book are those of the authors, but not necessarily of the publisher.
This book contains information sourced from authentic and highly regarded references, with reasonable efforts made to ensure the reliability of the data and information presented. The authors, editors, and publisher believe the information in this book to be accurate and true as of the date of publication. Every effort has been made to trace and credit the copyright holders of all materials included. However, the authors, editors, and publisher cannot assume responsibility for the validity of all materials or the consequences of their use. Should any copyright material be found unacknowledged, please inform the publisher so that corrections may be made in future reprints.

Table of Contents

Preface xiii

Chapter 1
Innovative Applications of 3D Printing in Precision Agriculture and Food Production 1
P. W. Ingle, Sanjivani College of Engineering, Kopargaon, India
Sharmila A., Raj Kumar Goel Institute of Technology, Ghaziabad, India
P William, Sanjivani University, Kopargaon, India
Sachin Vasant Chaudhari, Sanjivani College of Engineering, Kopargaon, India
N. Yogeesh, Government First Grade College, Tumkur, India
Firas Tayseer Ayasrah, Al Ain University, Al Ain, UAE

Chapter 2
Integration of IoT-Enabled 3D-Printed Sensors for Real-Time Smart Farming Applications 27
Rohith Bommalla Naresh, Astra Trading and Investment LLC, USA
Shafat Khan, Department of Computer Science, College of Computer Science, King Khalid University, Abha, Saudi Arabia
M. Arul, Department of Business Administration, Annamalai University, India
Vijayakumar Thota, NSB Bangalore, Bangalore University, India
S. G. Balaji, MEASI Institute of Management, Chennai, India
I. Yasar Shariff, Department of Information System and Management, The Quaide Milleth College for Men, Chennai, India

Chapter 3
Justice in Virtual Worlds: Ethical Dilemmas of AI and 3D Printing in the Metaverse 51
Zahid Parwez, Xavier Law School, XIM University, Odisha, India

Chapter 4
Green Agricultural Innovation: Combining 3D Printing, XR, and AI for Sustainable Farming 71
S. Seethalakshmi, Saveetha Institute of Medical and Technical Sciences, Chennai, India
Anju Mohan, Saveetha Institute of Medical and Technical Sciences, Chennai, India
U. Marimuthu, Saveetha Institute of Medical and Technical Sciences, Chennai, India
K. S. Alakumarimuthu, Government Arts College Melur, Madurai, India

Chapter 5
Human-Centric AI in Smart Farming: Paving the Way for Agriculture 5.0........ 97
Rajendrakumar A. Kapgate, Sanjivani College of Engineering, India
Vijayshree Khedkar, Symbiosis Institute of Technology, Pune, India
Ayushi Kapgate, Symbiosis Institute of Technology, Pune, India
P. M. Patare, Sanjivani College of Engineering, India
Michael Onyema Edeh, Coal City University, Enugu, Nigeria

Chapter 6
Enhancing Smart Agriculture Through Digital Twins: A Comprehensive Analysis........ 123
Dipesh B. Pardeshi, Sanjivani College of Engineering, Kopargaon, India
Prachi Lokhande, Sanjivani College of Engineering, Kopargaon, India
Pranali Gaikwad, Sanjivani College of Engineering, Kopargaon, India
Sanika Pandit, Sanjivani College of Engineering, Kopargaon, India
Mudassir Khan, King Khalid University, Saudi Arabia
Shashi Bhushan, Universiti Teknologi Petronas, Malayisa

Chapter 7
Comparative Evaluation of 3D Point Cloud Tools in Agriculture 5.0 Using Digital Twins 151
Rajendrakumar A. Kapgate, Sanjivani College of Engineering, India
Vijayshree Khedkar, Symbiosis Institute of Technology, India
Ayushi Kapgate, Symbiosis Institute of Technology, India
P. M. Patare, Sanjivani College of Engineering, India
Michael Onyema Edeh, Coal City University, Enugu, Nigeria

Chapter 8
Cybersecurity in Agriculture 5.0: Monitoring Agricultural Threats Through Digital Twin Technology 179
P. C. Senthil Mahesh, Excel Engineering College, India
Rohini Ganapathi, St..Joseph's Institute of Technology, India
Vivek Pandiya Raj, Vel Tech Rangarajan Dr. Sagunthala R&D Institute of Science and Technology, India
Natchiammai Palaniappan, Mookambigai College of Engineering, India
Sangeetha Kuppusamy, Kongu Engineering College, India
Manjunathan Alagarsamy, K. Ramakrishnan College of Technology, Trichy, India

Chapter 9
Improving Farm Management and Yield Prediction With Digital Twins in Precision Agriculture 203
K. Sona, Saveetha Institute of Medical and Technical Sciences, Chennai, India
S. Thangamayan, Saveetha Institute of Medical and Technical Sciences, Chennai, India
B. Lavaraju, Saveetha Institute of Medical and Technical Sciences, Chennai, India
K. S. Varsha, Saveetha Institute of Medical and Technical Sciences, Chennai, India
M. Raju, Saveetha Institute of Medical and Technical Sciences, Chennai, India
J. Christo Francis, Saveetha Institute of Medical and Technical Sciences, Chennai, India

Chapter 10
Next-Generation Farming: Digital Twins and Virtual Environments for Sustainable Agriculture 227
Sandeep Bhatia, Jamia Millia Islamia, New Delhi, India
Zainul Abdin Jaffery, Jamia Millia Islamia, New Delhi, India
Shabana Mehfuz, Jamia Millia Islamia, New Delhi, India
Neha Goel, Raj Kumar Goel Institute of Technology, India

Compilation of References 251

About the Contributors 267

Index 271

Detailed Table of Contents

Preface.. xiii

Chapter 1
Innovative Applications of 3D Printing in Precision Agriculture and Food Production 1
P. W. Ingle, Sanjivani College of Engineering, Kopargaon, India
Sharmila A., Raj Kumar Goel Institute of Technology, Ghaziabad, India
P William, Sanjivani University, Kopargaon, India
Sachin Vasant Chaudhari, Sanjivani College of Engineering, Kopargaon, India
N. Yogeesh, Government First Grade College, Tumkur, India
Firas Tayseer Ayasrah, Al Ain University, Al Ain, UAE

This study is to investigate how 3D printing is revolutionizing smart agriculture and food processing so that sustainability, efficiency, and product customizing may be increased. 3D printing allows customized equipment, components, and food products appropriate for particular nutritional and agricultural needs to be manufactured. This is practical when combining technologies including artificial intelligence (AI), the Internet of Things (IoT), and data analytics. The research looks at how 3D printing influences manufacturing efficiency, environmental sustainability, and economic performance under numerous application situations. Among other things, these circumstances include advances in smart sensors, accurate tools, and food processing. Data from real-world applications and performance assessments indicates major benefits including reduced production costs, greater material and energy economy, and more adaptability in response to changes in the market and the surroundings.

Chapter 2
Integration of IoT-Enabled 3D-Printed Sensors for Real-Time Smart Farming Applications 27
Rohith Bommalla Naresh, Astra Trading and Investment LLC, USA
Shafat Khan, Department of Computer Science, College of Computer Science, King Khalid University, Abha, Saudi Arabia
M. Arul, Department of Business Administration, Annamalai University, India
Vijayakumar Thota, NSB Bangalore, Bangalore University, India
S. G. Balaji, MEASI Institute of Management, Chennai, India
I. Yasar Shariff, Department of Information System and Management, The Quaide Milleth College for Men, Chennai, India

Agriculture is no longer bound by tradition—today, it's being transformed by innovation. Among the most groundbreaking shifts is the fusion of Internet of Things (IoT) technology with 3D-printed sensors. This paper explores how this powerful combination is helping farmers make smarter decisions, conserve precious resources, and grow more food sustainably. By placing custom-built sensors directly into the soil and connecting them to cloud platforms, we can now track moisture, temperature, and pH in real time. Our field experiments tested two different 3D-printed sensors alongside a traditional sensor across five farms. The results were compelling: Sensor A consistently outperformed the others in accuracy, reliability, and yield outcomes. Farms using these smart systems saw an 18% increase in crop yield and used 25% less water. More than just numbers, these results show the enormous potential of democratizing technology—making cutting-edge tools affordable and adaptable for every farmer, everywhere.

Chapter 3
Justice in Virtual Worlds: Ethical Dilemmas of AI and 3D Printing in the Metaverse 51
Zahid Parwez, Xavier Law School, XIM University, Odisha, India

Artificial intelligence (AI) use in court systems has altered their legal decision making and execution process. Major concerns of due process, 3D Printing, judicial ethics, openness, and public confidence are brought up by artificial intelligence. This is true even if these technologies provide consistency, efficiency, and data-driven insights. The difficult barriers of using artificial intelligence in the judicial system are investigated in this work. Fairness, algorithmic prejudice, 3D Printing, judicial power loss, and AI responsibility are discussed in this article. Stakeholder polls, empirical data analysis Indian and Latin American legal case studies were part of this research project. defendants are using technology more sparingly as they worry it may damage the judicial system. Studies on the frequency of mistakes, which begs ethical and pragmatic questions, indicate that artificial intelligence is unable to make high-stakes choices like bail and criminal punishment. The paper ends with several legislative suggestions for controlling artificial intelligence.

Chapter 4
Green Agricultural Innovation: Combining 3D Printing, XR, and AI for Sustainable Farming 71
S. Seethalakshmi, Saveetha Institute of Medical and Technical Sciences, Chennai, India
Anju Mohan, Saveetha Institute of Medical and Technical Sciences, Chennai, India
U. Marimuthu, Saveetha Institute of Medical and Technical Sciences, Chennai, India
K. S. Alakumarimuthu, Government Arts College Melur, Madurai, India

Combining 3D printing, Extended Reality (XR), and artificial intelligence (AI), the Green Agricultural Innovation offers a breakthrough approach to raise agricultural production and sustainability. The aim of this paper is to study the opportunities for cooperation across many technologies in the domains of resource optimization, waste reduction, and yield improvement. Potential means to improve sustainable farming operations include artificial intelligence (AI) for resource management and predictive analytics, augmented reality (XR) for immersive training and real-time simulations, and 3D printing for the production of unique agricultural goods. A quantitative and comparative research technique was used to assess the effects these technologies had on agricultural productivity and sustainability throughout a broad spectrum of farm kinds. According to the research, 3D, XR, and AI taken together provide the best sustainability and productivity increases. The study underlines the importance of combining many technical solutions to effectively achieve sustainable agriculture outcomes.

Chapter 5
Human-Centric AI in Smart Farming: Paving the Way for Agriculture 5.0........ 97
Rajendrakumar A. Kapgate, Sanjivani College of Engineering, India
Vijayshree Khedkar, Symbiosis Institute of Technology, Pune, India
Ayushi Kapgate, Symbiosis Institute of Technology, Pune, India
P. M. Patare, Sanjivani College of Engineering, India
Michael Onyema Edeh, Coal City University, Enugu, Nigeria

Agriculture 5.0 marks a new era in food production driven by advanced technologies and human-centered artificial intelligence. Unlike Agriculture 4.0, which focuses mainly on efficiency and innovation, Agriculture 5.0 promotes collaboration where AI supports human decision-making. This study explores the ethical, social, and environmental impacts of human-centric AI in smart farming. Through case studies, statistical analysis, and comparative research, we show how open, inclusive, and sustainable AI can enhance agricultural productivity, efficiency, and farmer satisfaction. Task-based evaluations and ANOVA tests highlight the advantages of human-focused AI over traditional models. Beyond technical gains, the paper addresses ethical dilemmas, digital inequality, and data bias, proposing a cooperative model for agricultural tech. Embracing this human-first approach is vital for creating smart, fair, and sustainable agriculture.

Chapter 6
Enhancing Smart Agriculture Through Digital Twins: A Comprehensive Analysis 123
Dipesh B. Pardeshi, Sanjivani College of Engineering, Kopargaon, India
Prachi Lokhande, Sanjivani College of Engineering, Kopargaon, India
Pranali Gaikwad, Sanjivani College of Engineering, Kopargaon, India
Sanika Pandit, Sanjivani College of Engineering, Kopargaon, India
Mudassir Khan, King Khalid University, Saudi Arabia
Shashi Bhushan, Universiti Teknologi Petronas, Malayisa

One of the biggest advances in smart agriculture is digital technology mixed with precision farming. Digital Twins (DTs), virtual representations of physical systems that enable real-time monitoring, modeling, and process optimization, have revolutionized agriculture. This article examines how distributed technologies (DTs), cloud computing, IoT devices, and AI-driven analytics might enhance agricultural productivity, resource efficiency, and farm management. Statistics utilizing analysis of variance (ANOVA) and post-hoc testing show that digital twin-based farming is superior than traditional and IoT-based farming in production consistency and environmental responsiveness. When sensor data resolution was increased by 5900% and system variability was decreased, all key crop types demonstrated 20% to 36% output gains. Histograms, violin plots, and boxplots demonstrate DT systems' accuracy and stability. The study shows that digital twins enhance sustainable agriculture.

Chapter 7
Comparative Evaluation of 3D Point Cloud Tools in Agriculture 5.0 Using Digital Twins 151
Rajendrakumar A. Kapgate, Sanjivani College of Engineering, India
Vijayshree Khedkar, Symbiosis Institute of Technology, India
Ayushi Kapgate, Symbiosis Institute of Technology, India
P. M. Patare, Sanjivani College of Engineering, India
Michael Onyema Edeh, Coal City University, Enugu, Nigeria

Agriculture 5.0 marks a shift in farming through technologies like digital twins, AI, robotics, and real-time 3D spatial modeling. A key innovation is the use of 3D point cloud methods for accurate virtual modeling of agricultural environments.

This study compares eight 3D reconstruction tools—Agisoft, Pix4DMappers, ODM, DJi Terra, Meshroom, Regard3D, Reality Capture, and 3DF Zephyr—based on geometric integrity in five shape metrics: sphericity, planarity, linearity, anisotropy, and curvature. Using photogrammetry datasets, shape-index scoring, and regression analysis, we evaluated their suitability for digital twin use. Results show no single tool excels in all aspects; software choice should depend on specific goals. Agisoft scored high in curvature and sphericity, ODM performed best overall, and Meshroom and Regard3D suited high-anisotropy designs. The study highlights the importance of volumetric and curvature accuracy in biological modeling and promotes open-source tools for cost-effective digital twin deployment in precision farming.

Chapter 8
Cybersecurity in Agriculture 5.0: Monitoring Agricultural Threats Through Digital Twin Technology 179
P. C. Senthil Mahesh, Excel Engineering College, India
Rohini Ganapathi, St..Joseph's Institute of Technology, India
Vivek Pandiya Raj, Vel Tech Rangarajan Dr. Sagunthala R&D Institute of Science and Technology, India
Natchiammai Palaniappan, Mookambigai College of Engineering, India
Sangeetha Kuppusamy, Kongu Engineering College, India
Manjunathan Alagarsamy, K. Ramakrishnan College of Technology, Trichy, India

Within the framework of the digital revolution in the agriculture industry, Agriculture 5.0 is a noteworthy accomplishment reflecting a major progress. The agriculture industry employs a broad spectrum of technologies to reach this aim: digital twin systems, big data, the Internet of Things (IoT), and artificial intelligence (AI). Investigating the use of digital twins with the aim of monitoring and enhancing agricultural cybersecurity is the main goal of this work. Real-time simulations of physical systems are examined in this study along with potential uses for these simulations to reduce risks in cyber-physical environments. The growing reliance of smart farming on networked equipment, which calls for the use of strict and easily available security techniques, has changed the risk landscape.

Chapter 9
Improving Farm Management and Yield Prediction With Digital Twins in Precision Agriculture 203
K. Sona, Saveetha Institute of Medical and Technical Sciences, Chennai, India
S. Thangamayan, Saveetha Institute of Medical and Technical Sciences, Chennai, India
B. Lavaraju, Saveetha Institute of Medical and Technical Sciences, Chennai, India
K. S. Varsha, Saveetha Institute of Medical and Technical Sciences, Chennai, India
M. Raju, Saveetha Institute of Medical and Technical Sciences, Chennai, India
J. Christo Francis, Saveetha Institute of Medical and Technical Sciences, Chennai, India

Using digital twin technology into precision farming is a significant first toward the development of agricultural techniques more intelligent and less harmful to the environment. Basically, virtual copies of real-time data from sensors, satellites, and artificial intelligence models, digital twins aim to copy, monitor, and enhance agricultural activity by means of synchronizing real-time data. This paper investigates utilizing a comprehensive literature review, simulated trials, and statistical validation using analysis of variance (ANOVA) the possibilities of digital twins to enhance farm management and production prediction. The results reveal in terms of agricultural productivity, resource efficiency, and prediction accuracy that digital twin systems outperform both traditional techniques and solutions based on the Internet of Things. Notable gains also came in important performance measures like water use, decision-making time required, and input cost savings. The results underscore the importance of digital twins driven by data in the agricultural field.

Chapter 10
Next-Generation Farming: Digital Twins and Virtual Environments for Sustainable Agriculture.. 227
Sandeep Bhatia, Jamia Millia Islamia, New Delhi, India
Zainul Abdin Jaffery, Jamia Millia Islamia, New Delhi, India
Shabana Mehfuz, Jamia Millia Islamia, New Delhi, India
Neha Goel, Raj Kumar Goel Institute of Technology, India

- Digital technologies are transforming agriculture, provide innovative solutions for sustainability, efficiency, and resilience in farming. Among these technologies, integration of Digital Twins (DTs) and Virtual Environments (VEs) is revolutionizing agriculture by optimizing agricultural processes, reducing resource consumption, and enhancing productivity. Digital Twins create real-time, data-driven replicas of crops, soil, livestock, and farming infrastructure, enabling continuous monitoring and predictive analytics. These virtual models help farmers make data-informed decisions, improving resource efficiency and minimizing environmental impact. Virtual Environments, including Virtual Reality (VR) and Augmented Reality (AR), complement Digital Twins by providing immersive simulations for precision farming, training, and farm management.

Compilation of References 251

About the Contributors 267

Index 271

Preface

The world of agriculture is undergoing an evolution driven by the rapid convergence of advanced technologies such as 3D printing, Artificial Intelligence (AI), the Internet of Things (IoT), Digital Twins, and immersive virtual environments. As the agricultural sector transitions from traditional practices to data-driven and precision-based models, there arises a pressing need to explore how these emerging technologies can be thoughtfully integrated to promote sustainability, improve productivity, and ensure food security for a growing global population.

This edited volume, *3D Printing and Digital Twins in Smart Agriculture*, brings together researchers, technologists, and scholars from across the globe to examine this pivotal shift. The chapters explore how cutting-edge technologies are revolutionizing various facets of agriculture from equipment customization and sensor-based monitoring to ethical concerns in virtual systems and cybersecurity in the era of Agriculture 5.0.

The diversity of perspectives represented in this book reflects the interdisciplinary nature of modern agriculture. Each chapter offers valuable insights into current developments, practical applications, and future possibilities of combining digital innovation with agricultural science. Together, they present a forward-looking vision of how agriculture can meet the demands of the future through smart, ethical, and sustainable solutions.

CHAPTER OVERVIEW

Chapter 1: Innovative Applications of 3D Printing in Precision Agriculture and Food Production

This chapter explores the transformative impact of 3D printing on equipment manufacturing, food customization, and agricultural efficiency. It highlights real-world use cases demonstrating cost savings, sustainability improvements, and enhanced adaptability.

Chapter 2: Integration of IoT-Enabled 3D-Printed Sensors for Real-Time Smart Farming Applications

The authors present a practical study on the effectiveness of 3D-printed IoT sensors in tracking soil conditions, leading to improved crop yields and water conservation. Field experiments reveal the benefits of integrating affordable, custom sensors into smart farming systems.

Chapter 3: Justice in Virtual Worlds – Ethical Dilemmas of AI and 3D Printing in the Metaverse

This chapter critically examines the legal and ethical challenges surrounding the use of AI and 3D printing in virtual environments, especially within judicial contexts. It discusses algorithmic bias, transparency, and the implications for trust in digital systems.

Chapter 4: Green Agricultural Innovation Combining 3D Printing, XR, and AI for Sustainable Farming

Focusing on technological convergence, this chapter investigates how XR (Extended Reality), AI, and 3D printing can collectively drive resource efficiency and immersive training in sustainable agriculture.

Chapter 5: Human-Centric AI in Smart Farming – Paving the Way for Agriculture 5.0

This contribution advocates for a human-first approach to AI in farming. It examines the ethical, social, and practical dimensions of human-centric AI, emphasizing inclusivity, transparency, and collaboration in agricultural decision-making.

Chapter 6: Enhancing Smart Agriculture Through Digital Twins – A Comprehensive Analysis

A data-rich analysis of how Digital Twins, cloud computing, and AI analytics can optimize farm performance. The chapter includes comparative data visualizations and statistical tests that highlight the advantages of this approach over traditional methods.

Chapter 7: Comparative Evaluation of 3D Point Cloud Tools in Agriculture 5.0 Using Digital Twins

This chapter presents a comparative evaluation of various 3D reconstruction tools used in Digital Twin modeling. It provides key performance metrics and software recommendations based on specific agricultural modeling needs.

Chapter 8: Cybersecurity in Agriculture 5.0 – Monitoring Agricultural Threats Through Digital Twin Technology

With the rise of connected farming devices, cybersecurity becomes paramount. This chapter discusses how Digital Twin simulations can detect, monitor, and mitigate cyber threats in modern agricultural systems.

Chapter 9: Improving Farm Management and Yield Prediction with Digital Twins in Precision Agriculture

The authors explore how integrating Digital Twins with real-time data enhances yield forecasting and farm management. Statistical analyses validate the performance improvements in decision-making, resource efficiency, and cost management.

Chapter 10: Next-Generation Farming – Digital Twins and Virtual Environments for Sustainable Agriculture

The final chapter expands the discussion to Virtual Reality (VR) and Augmented Reality (AR) as complementary technologies to Digital Twins. It emphasizes their role in immersive training, predictive modeling, and sustainable farming practices.

Through a blend of empirical research, case studies, and theoretical analysis, this book offers an essential resource for academics, practitioners, and policymakers interested in shaping the future of agriculture through innovation. It is our hope that the insights herein will inspire further exploration and responsible implementation of these groundbreaking technologies in agricultural systems around the world.

Editors,

P. William
Sanjivani University, India

Sharmila A.
Raj Kumar Goel Institute of Technology, India

A. K. Haghi

University of Coimbra, Portugal

Chapter 1
Innovative Applications of 3D Printing in Precision Agriculture and Food Production

P. W. Ingle
https://orcid.org/0009-0005-5956-7866
Sanjivani College of Engineering, Kopargaon, India

Sachin Vasant Chaudhari
https://orcid.org/0009-0005-8856-8905
Sanjivani College of Engineering, Kopargaon, India

Sharmila A.
https://orcid.org/0000-0002-9871-3871
Raj Kumar Goel Institute of Technology, Ghaziabad, India

N. Yogeesh
https://orcid.org/0000-0001-8080-7821
Government First Grade College, Tumkur, India

P William
https://orcid.org/0000-0002-0610-0390
Sanjivani University, Kopargaon, India

Firas Tayseer Ayasrah
https://orcid.org/0000-0002-6477-000X
Al Ain University, Al Ain, UAE

ABSTRACT

This study is to investigate how 3D printing is revolutionizing smart agriculture and food processing so that sustainability, efficiency, and product customizing may be increased. 3D printing allows customized equipment, components, and food products appropriate for particular nutritional and agricultural needs to be manufactured. This is practical when combining technologies including artificial intelligence (AI), the Internet of Things (IoT), and data analytics. The research looks at how 3D printing influences manufacturing efficiency, environmental sustainability, and economic performance under numerous application situations. Among other things, these circumstances include advances in smart sensors, accurate tools, and

DOI: 10.4018/979-8-3373-7077-4.ch001

ght © 2026, IGI Global Scientific Publishing. Copying or distributing in print or electronic forms without written permission of IGI Global Scientific Publishing is prohibited. Use of this chapter to generative artificial intelligence (AI) technologies is expressly prohibited. The publisher reserves all rights to license its use for generative AI training and machine learning model development.

food processing. Data from real-world applications and performance assessments indicates major benefits including reduced production costs, greater material and energy economy, and more adaptability in response to changes in the market and the surroundings.

1. INTRODUCTION

Agriculture, as we know it, is changing. No longer just a practice rooted in soil and sunshine, farming has evolved into a dynamic field of data, design, and digital innovation. Across the world, farmers, food producers, and researchers are facing new and complex challenges—climate unpredictability, resource scarcity, growing food demand, and the urgent need for sustainability. In response, a surprising technology once confined to industrial labs is now making its way into the fields and kitchens of tomorrow: 3D printing. Also known as additive manufacturing, 3D printing creates physical objects layer by layer from digital designs. This process has already transformed industries like aerospace and medicine, and now, it's beginning to reshape how we grow food, care for crops, and even how we eat. Unlike mass manufacturing, 3D printing allows for quick customization, minimal waste, and the flexibility to adapt tools and products to highly specific needs. These characteristics make it a perfect match for the unique demands of precision agriculture and sustainable food production. The transformation starts small—literally—with 3D-printed sensors. These include soil moisture sensors, temperature gauges, and nutrient probes. As shown in Figure 1, these printed devices can be customized to the unique layout of a farm or the structure of a greenhouse. They collect detailed data about the environment and send that data through IoT networks, communicating with central systems through low-power wireless protocols like LoRa and ZigBee. From there, the data reaches cloud-based platforms where machine learning algorithms and AI go to work—analyzing patterns, predicting crop needs, and helping farmers make informed decisions (Derossi, Caporizzi, Azzollini, & Severini, 2018). But this story goes far beyond sensors. In this introduction, we'll explore how 3D printing is enabling farmers to create custom tools, personalize crop care, build smart infrastructure, and even print food. We'll also look at how this technology connects seamlessly with other innovations like AI and IoT, forming a smarter, faster, and more sustainable agricultural ecosystem.

Printing the Future, One Tool at a Time

Every farmer knows the frustration of a broken part at the worst possible moment—an irrigation nozzle cracks mid-season or a sensor mount gives out just before harvest. Replacements can take days or weeks to arrive, and in rural or remote areas, they may not be available at all. This is where 3D printing steps in. With just a printer and a digital file, farmers can print their own replacement parts on demand—often at a fraction of the cost and time (Camille et al., 2017). But it's not just about replacing what's broken. Farmers are now using 3D printing to create entirely new tools that suit their specific soil types, crops, or weather conditions. Need a narrow hoe for tightly spaced plants? Or a sprinkler that rotates just the right way? With basic design software and a desktop printer, these ideas can become reality in just hours. This level of customization empowers small farms to innovate locally and helps larger operations adapt faster. Even educational institutions and research farms are using 3D printers to prototype tools and conduct field experiments with quick turnaround.

Figure 1. Integrated precision agriculture system

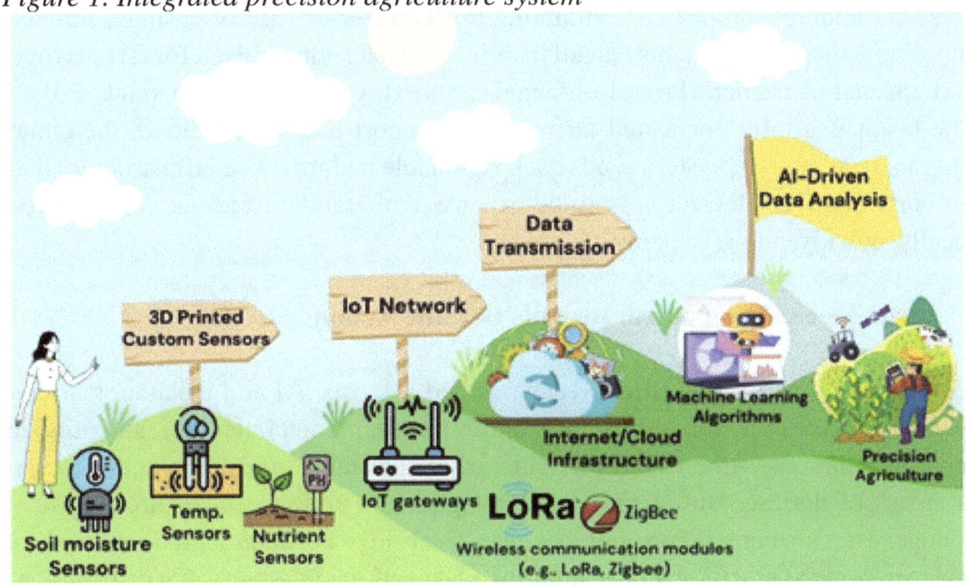

Smart Sensors for Smarter Farming

Precision agriculture relies heavily on real-time data—and that data starts at the ground level. With 3D printing, researchers and engineers are now designing affordable, field-ready sensor housings that can be printed locally, adjusted for durability, and easily replaced when needed. These sensors—monitoring moisture, temperature, pH, and more—play a vital role in helping farmers respond quickly to changes in the environment. And because the components are often printed using biodegradable or recyclable materials, they're not only efficient—they're sustainable too (Podchasoy, 2021). What's more, many open-source sensor designs are now available online, meaning a smallholder farmer with access to a community printer and basic training can build their own smart monitoring system. It's a step toward democratizing agricultural technology—putting innovation into the hands of those who need it most.

From Field to Cloud: The Role of IoT

Sensors are only useful if they can send data where it needs to go. That's where Internet of Things (IoT) networks come in. These networks connect field devices to cloud platforms via small, low-power gateways that use wireless communication like LoRa and ZigBee. The brilliance of 3D printing here is its ability to build and adapt these communication devices. Mounting brackets, weatherproof casings, antenna supports—these components can all be printed on site, customized for farm layout, and updated as needed (Fred, Lohrengel, Neubert, Camila, & Czelusniak, 2014). The result is a fully connected farm: sensors report data to the cloud, the cloud runs analysis, and the system sends back actionable insights. The infrastructure that supports this flow doesn't have to be expensive or standardized—it can be made locally, quickly, and affordably.

AI Meets Agriculture: Insights Become Action

Once the data from the field reaches cloud systems, AI and machine learning algorithms take over. These systems identify trends, predict outcomes, and provide recommendations—like when to irrigate, what nutrient to apply, or whether a crop is at risk of disease. But turning those insights into action still requires physical change. That's where 3D printing becomes the bridge between digital intelligence and real-world application. For example, if AI detects inconsistent water distribution, a farmer can design and print a new irrigation head optimized for the problem area. If an unusual pest pattern emerges, a printed trap with a specialized design can be produced in real time (Ahmed, 2023). This closed-loop system—where sensors gather data, AI interprets it, and 3D printers create the physical solution—represents a major evolution in farming efficiency. It's fast, flexible, and it keeps the power to innovate in the hands of the farmer.

Food Innovation: Printing What We Eat

3D printing isn't just revolutionizing farm tools—it's also changing what happens to the food after it's grown. In food labs and research kitchens around the world, innovators are using 3D printing to create customized, nutritious, and sustainable meals. This means printing food using pastes made from fruits, vegetables, legumes, and even protein-rich byproducts. The goal? Reduce food waste, improve nutritional precision, and open up new avenues for plant-based eating. In schools, hospitals, and elder-care settings, this approach offers personalized meals that meet dietary needs and look appealing too. One particularly exciting development is the use of agricultural byproducts—things like fruit pulp, vegetable peels, or underused grains—to create food-safe inks for printing. These materials are abundant, low-cost, and nutritionally valuable. When turned into printable matter, they extend the life of harvests and reduce overall waste.

Building Infrastructure with Sustainability in Mind

As urban agriculture grows and more farms operate in indoor or vertical environments, there's a need for customized infrastructure—racks, trays, nutrient channels, lighting mounts, and more. 3D printing is perfect for this, offering the flexibility to print structures that fit tight spaces and support creative growing systems. Some projects are even experimenting with biodegradable materials made from crop residues—using husks, straw, or fiber waste to produce the printing filament itself. This closed-loop model—where farms produce the materials they need to build their own systems—is not just efficient, it's circular, sustainable, and remarkably low-waste.

A Technology That Puts Innovation in Every Hand

Perhaps the most powerful part of 3D printing in agriculture is its accessibility. Once considered high-tech and unreachable, desktop printers are now affordable and easy to use. Digital designs can be shared freely, improved locally, and printed globally. This changes the power dynamics in food production. Farmers are no longer just end-users—they can become creators. Students can learn by building their own tools. Communities can solve their own equipment problems. Innovation, once centralized, is now becoming personal and participatory.

The story of 3D printing in precision agriculture is just beginning, but its potential is already clear. From the first printed sensor to the last layer of a customized protein bar, this technology offers a new way to build, grow, and eat. It brings together some of the most important goals of modern farming: efficiency, sustainability, customization, and resilience. Whether it's a family farm printing its own irrigation tools, a smart greenhouse adjusting its nutrient delivery based on sensor feedback, or a food startup creating meals tailored to health needs—3D printing is quietly, layer by layer, reshaping how we grow and nourish. And most importantly, it's not a distant dream. It's already in motion. All it takes is imagination, a digital file, and a printer.

2. REVIEW OF LITERATURE

The development of additive manufacturing, which is also known as three-dimensional printing, is an incredible innovation that has numerous applications in the fields of intelligent agriculture and food processing. In addition to this, it offers real-time data on the status of plants, the levels of nutrients, and the conditions of the soil, as well as adaptive farming tools and autonomous agricultural instruments such as drones and robots (Hamilton, Alici, & Marc, 2017). The introduction of 3D printing has brought about a profound transformation in the food processing business. This is because it has made it possible to provide individualized nutrition solutions, particularly in the field of therapeutic nutrition. In addition to this, it makes it possible to alter the structures and textures of food, which ultimately contributes to the creation of one-of-a-kind sensory experiences and superior goods. The use of a variety of protein sources and the reduction of food waste by ten to thirty percent are two of the ways in which 3D printing contributes to the production of food in a manner that is more environmentally friendly. The researchers discovered that by using artificial intelligence and sensors that are enabled by the Internet of Things in three dimensions, it is possible to increase agricultural yields by ten to fifteen percent while simultaneously drastically reducing crop damage. In addition to this, they have the potential to contribute to a twenty percent decrease in the overall amount of power that is used, a twenty to thirty percent reduction in the amount of labor that is required, and a twenty to twenty-five percent reduction in the amount of water that is used. Despite this, there are a few obstacles that prevent widespread adoption. Costly pricing, advanced technical and design skills, and limitations on the rate and volume of manufacture are some of the obstacles that must be overcome throughout the manufacturing process. In addition, it is necessary to address concerns pertaining to the laws and the safety of the area. As a result of recent advancements in bioprinting, multifunctional materials, blockchain technology, and artificial intelligence, the future of three-dimensional printing seems to be bright for a variety of businesses. These developments may, in the long run, be able to expand the possibilities of 3D printing, which would then make it possible to produce things of a better quality, ways of operation that are more efficient, and operations that are less harmful to the environment.

Beginning with the first hand plow and continuing all the way up to the most cutting-edge soil analyzers that make use of satellite technology, innovation has always been at the core of farming. To remain up with the times and the rising number of issues that are being tackled, farmers and scientists are always exploring for strategies to make farming smarter, more sustainable, and more successful. Because of this forward-thinking tendency, the technology known as 3D printing, which was traditionally exclusively utilized in engineering workshops and aerospace labs, is now being used in contemporary greenhouses, kitchens, and even dining rooms. This is a significant change from the past. The process of 3D printing is straightforward, yet it yields tangible results by stacking digital concepts on top of one another. Because of this, it is of such high quality. Not only does this ability make things simpler, but it is also causing a shift in the way things are done in the agricultural industry. As part of their most recent study, researchers investigated the ways in which three-dimensional printing is being used across a wide range of various businesses. Precision farming, the customization of equipment, and the production of food were their primary focuses. A field that is experiencing issues ranging from a lack of manpower to environmental instability, which links all of these different sectors, is seeing an alarming rate of expansion in the requirements for speed, flexibility, and cost-effectiveness. This is a field that is experiencing these challenges. First, let's take a look at the application that provides the most evident and practical benefits: the customization of agricultural machinery and components. It is likely that those who are familiar with agricultural machinery are aware of how excruciating it can be to have a damaged component when there is a high demand for it. There is a possibility that it may take many days or even weeks for replacements to arrive at your location, particularly if you reside in a rural area that is some distance away. This is the reason why the use of 3D printing could be helpful. Consider the scenario in which a farmer has an accident that results in the destruction of a pipe that is essential for an irrigation system or a drone attachment. Something like this would be a terrible thing to take place. The fact that they can print out a digital blueprint allows them to create a new component on the farm in a matter of hours, so they do not have to wait. Because of this, the process is sped up. Therefore, there is no longer a need for individuals to engage in waiting. There is a large number of research that demonstrates how the same strategy has resulted in cost and time savings. The use of 3D printers in agriculture is becoming more crucial due to the fact that these machines are capable of producing a wide variety of components, ranging from sprayer nozzles to particular components for harvesting equipment.

Additionally, additive manufacturing, which is also referred to as three-dimensional printing in certain circles, is already being used in a wide variety of applications. The purpose of this work is to offer a concise overview of the many applications of 3D printing in the fields of agriculture, food processing, and environmental monitoring and preservation. The many types of materials that are used in the manufacture of

3D printed components, the various printing procedures that are used throughout the process, and the potential applications of these materials in the future are the primary topics that are covered in this research. PLA, also known as polylactic acid, and ABS thermoplastics are the materials that are used for 3D printing in the agricultural sector the most often. This is mostly due to the fact that the manufacture of these materials is both straightforward and inexpensive (Dizon et al., 2019). Through the use of direct extrusion, individuals who have difficulty swallowing are able to personalize their diet and consume a greater quantity of food. When environmental considerations are taken into consideration, this research investigates a wide range of primary uses for three-dimensional printing. Desalination of water and monitoring of air quality are two examples of different applications for these technologies. There is a possibility that the advancement of three-dimensional printing may entirely revolutionize the procedures that are used in the manufacturing and formulation of food goods. The development of meals that include a wide variety of ingredients and additions has been essential ever since the beginning of time in order to make foods that have distinctive sensory properties. The capacity of additive manufacturing to deliver knowledge that cannot be obtained in other technologies is one of the primary factors contributing to the success of additive manufacturing in industries such as food processing. In order to do this, the purpose of this article is to offer a summary of the current level of knowledge on the many features of printing material sources and the way in which these components influence printing processes. In addition, this article provides further information by providing a comprehensive summary of the globalization of applications for food packaging, customized nutrition, and meals that have been printed on specific materials. To demonstrate the wide range of applications that may be found for 3D printing in the food industry, this is done. It is important to note that the basic challenges associated with 3D food printing have been overcome, which highlights the need of doing future research on this topic.

We are keeping a close check on the state of the ecology and the soil, which is another significant shift that is taking place right in front of our own eyes. All of the processes that take place in the soil must be understood in order to practice precision agriculture. It is possible that ordinary equipment will not function properly due to the fact that there are several varieties of crops, field layouts, and weather conditions. What would happen, however, if you were able to construct a probe or sensor housing that was specifically designed for the kind of soil and crop that you are working with? In order to accomplish this goal, researchers and early adopters are making extensive use of 3D printing. Data collection in the field is undergoing a transformation as a result of the development of equipment that is compact, practical, and inexpensive. Cornfields that have moisture sensors and vertical plants that have nutrient-measuring devices are two examples of the kind of technology that falls under this category. Currently, 3D printing is being exploited not only for the production of equipment but also for the production of food (Padhiary et al., 2023). That is, the dinner itself. The question of whether or not meals can be printed with 3D printing is now one of the most contentious issues in the world of contemporary literature. Imagine for a moment of a printer had the ability to transform lentils or peas into layers of protein-rich paste, so producing a snack that was not only beneficial to your health but also beautifully crafted. This concept is particularly significant at the present time since people all over the globe are increasingly leaning toward diets that are more plant-based and that persist for a longer period of time. Manufacturing food out of agricultural waste, such as grains, fruit dregs, and vegetable pulps, is becoming more common among research institutes and food technology companies. These 3D printers are being used to create food. The recycling of items that would otherwise be discarded is not only an inventive approach to recycle them, but it is also beneficial to the protection of the environment. Additionally, it provides new opportunities to personalize one's diet by making use of a variety of materials.

Through the use of contemporary technology, precision agriculture seeks to boost agricultural production while simultaneously reducing adverse impacts on the surrounding environment. Precision agriculture is a farming strategy that significantly boosts crop yields while simultaneously lessening the amount of waste produced. This is accomplished via the use of data analysis and contemporary technology. The agricultural practice that is often referred to as "precision agriculture" is one example of this. An approach that is pragmatic is required in order to find solutions to some of the most significant problems that contemporary agriculture is now confronted with. One of these problems is the challenge of feeding the growing population of the globe while simultaneously reducing the negative effect on the environment from agriculture. This article discusses two of the most recent advancements in precision agriculture: big data and the Internet of Things (IoT). Both of these innovations are discussed in detail. In the industry, both of these technologies are at the forefront of the competition. There are extensive reasons for these two most recent instances (Nyman & Sarlin, 2014), which are both examples. The objective of this review article is to present a comprehensive overview of the most recent advancements in precision and smart farming, as well as the challenges that these breakthroughs bring about and the potential applications of these findings. Specifically, this article takes a look at the present state of precision agriculture, paying particular attention to the most recent technological breakthroughs, such as sensors, drones, and machine learning. In addition, the article discusses some of the most significant challenges that precision agriculture must overcome, including the administration of data, the cost-effectiveness of technology advancements, and the acceptability of technological advancements.

Moreover, the fields of education and research are undergoing a transformation that is not quite as significant. In order to give students with new opportunities to learn via hands-on experience, agricultural colleges and innovation centers are using 3D-printed replicas of crop structures, animal organs, and even actual farm settings. As a means of producing these duplicates, many technologies are being used. Through the use of printed pest models conducted in interactive workshops, farmers have the opportunity to discover early infestations without putting their actual crops at risk. In addition to assisting individuals in acquiring knowledge more quickly, these models are also assisting in the process of bridging the gap between what individuals know about technology and how they can make use of it. In the area of controlled-environment agriculture (CEA), which encompasses hydroponics, aeroponics, and vertical farming, 3D printing is contributing to the construction of the infrastructure that will be used in the future. In contrast to more traditional farming methods, these systems are often used in indoor spaces that are rather tiny and need specialized components that have been manufactured specifically for them.

The employment of printed modules allows for the distribution of nutrients, the construction of support structures for lighting, and the holding of roots in order to provide growing conditions that are precise and efficient. Computers are used in the production of these components, which means that they may be improved and altered over time to function more effectively without wasting resources or requiring a significant amount of manufacturing. Additionally, this indicates that there is no need for mass production. Because of advancements in digital technology, this is now a real possibility. When people speak about the future of farming, sustainability will be a huge subject of discourse, and 3D printing will absolutely be a part of that debate. Among the studies that have been conducted in recent years, the one that investigated the process of converting waste into money has been one of the most fascinating. In order to develop biopolymers composed of agricultural waste, such as rice straw or maize husks, for use as printer filaments, researchers are now investigating and developing these materials. Farmers now have the ability to print everything from tool parts to packaging thanks to this technology. In addition to this, it prevents rubbish from being burned or allowed to fester. Now that this strategy has been put into effect, the cycle has come to an important end. In addition, additive manufacturing generates less waste than traditional methods since it constructs things layer by layer, utilizing just the components that are required for the creation of the final product.

Three-dimensional printing is an example of a cutting-edge technology that has the potential to boost the environmental sustainability and consumer attractiveness of existing food value chains. This includes the capacity to print in three dimensions. It may be possible to achieve this goal with the assistance of programs such as automated meal customization, on-demand food manufacturing, and regulations that reduce food waste. It is possible that the incorporation of nutrient-dense substrates, probiotics, bioactive compounds, and functional components into meals that have been meticulously prepared might assist to alleviate food shortages in countries where available fresh foods at reasonable prices are scarce. In an effort to improve the autonomy and effectiveness of their manufacturing processes, food processing firms all around the globe are experimenting with the use of 3D printing for food. There is an estimate that this technology will see a compound annual growth rate of around 55 percent. The purpose of this review article is to give a comprehensive examination of the technology (ASTM Committee F42 on Additive Manufacturing Technologies, 2012), beginning with the several methods and printers that are now available for use in the field of three-dimensional printing. In addition to the essential raw components required for a variety of printing applications, this coverage also include future precision and personalized food components. Within the scope of our research, we investigate the potential applications and prospects of three-dimensional food printing in the years to come, as well as the positive and negative aspects of the technologies involved.

To put it another way, 3D printing is not a foolproof method of solving problems. However, it is impossible to sidestep the issues that are brought up in the literature. It may be difficult to get materials that meet the standards of an industrial setting, and not all farms possess the digital skills necessary to manufacture components. On the other hand, there are still high-end printers that impose a significant financial burden. In contrast, there is a great deal of optimism to be found. The challenges that have been preventing development from progressing are gradually disappearing as a result of the proliferation of open-source communities, shared design libraries, and innovation centers in rural areas. Extension initiatives funded by universities and farmer cooperatives are starting to exchange training and materials in order to facilitate local innovation that is both effective and inexpensive. The combination of smart systems with 3D printing has been found to be an intriguing trend that requires more investigation, according to a significant amount of research. When you consider the following: a soil moisture sensor generates an indication and then transmits it to a farmer. Following an analysis of the data, a system that makes use of artificial intelligence will provide recommendations on modifications to the irrigation schedule as well as the design of the water distribution nozzle. After receiving approval from the farmer, the component is manufactured using a 3D printer that is connected to the farmer's manufacturing system. The process of setting it up and using it just takes a few hours. It is possible that the future of ultra-precision farming will include the implementation of these closed-loop, autonomous feedback systems. These systems make use of data to make changes to the physical environment. The ability of this new technology to provide farmers with the ability to immediately generate fresh ideas is one of the finest things about it. There are now other solutions available than large corporations and extensive supply networks. Either you can figure out how to fix the issue on your own or you may seek assistance from someone who lives on the other side of the planet and has already found a solution to the problem that you are experiencing. On the other hand, it not only transforms the farmer into a maker, designer, and innovator, but it also transforms the farmer into someone who is only concerned with technology.

3. 3D-PRINTED SENSORS AND IoT DEVICES FOR SMART FARMING

When farming was done in the past, intuition and observation were the most important factors. For example, farmers would observe how the leaves curled, feel how dry the soil was, or see how quickly the plants took off. Nevertheless, in the modern world, it is no longer sufficient. There is a need for assistance in contemporary farming because of the shifting weather patterns, the growing prices, and the need to feed more people than ever before. Additionally, this assistance is coming from a variety of sources that were not anticipated, such as 3D printers and smart gadgets. Upon first inspection, the use of 3D printing in farming does not seem to be a logical fit. Most of the time, it is connected to high-tech laboratories or mechanical components. However, it has come to light that this technology, which can construct items layer by layer from a digital design, is providing farmers with new methods to produce tools, monitor crops, and make more informed decisions—right from the beginning. Additionally, when combined with the Internet of Things (IoT), it increases its level of power even further. Imagine there is a farm where the earth itself communicates with the farmer to let them know when it needs water. Or a place where the temperature, humidity, and nutrient levels of the air are continuously measured, not by prohibitively costly instruments imported from other countries, but by minuscule tools that are manufactured directly on the farm. The Internet of Things (IoT) technologies and 3D-printed sensors have made this possibility a reality.

Figure 2. Using AI and image processing to monitor crop health

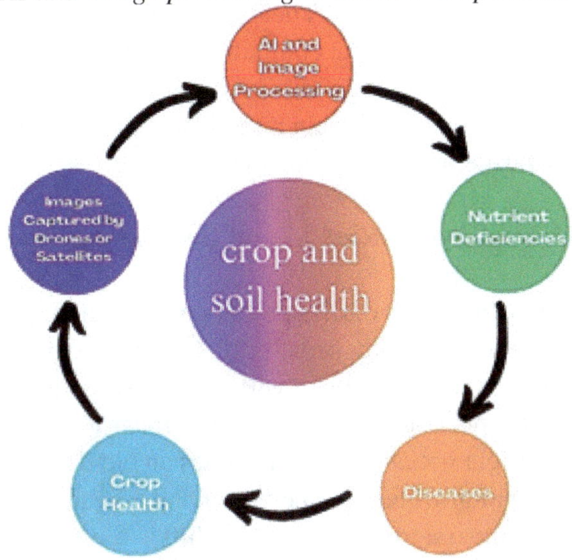

Making Tools that Fit the Farm

Not all farms are the same. Some grow leafy greens in rooftop gardens, others stretch across dry plains or steep hills. Yet most sensors and tools available on the market are one-size-fits-all. They're often too expensive, too bulky, or just not suited to the task. That's where 3D printing steps in. With just a printer, a computer file, and some basic material, a farmer can now create a custom-fit sensor in a matter of hours. Need a small moisture probe to stick between lettuce rows? Or a sensor housing that can handle heavy rain? Just design it—or download it from an open-source library—and print it. The result? Tools that actually fit the farmer's needs. And because they're printed from affordable, often biodegradable plastic, they're both low-cost and environmentally friendly. If something breaks, it can be reprinted. If something changes, it can be redesigned.

Making Smart Sensors Smarter

Of course, a printed sensor by itself can't do much. The real magic happens when it becomes smart—when it's connected to a network and starts sending data. That's where IoT comes in. Using tiny chips and low-energy communication systems like LoRa and ZigBee, these printed sensors can send information wirelessly across fields, orchards, or greenhouses. They don't need cell towers or complex infrastructure. They just quietly collect and transmit data: “The soil is dry here.” “The temperature has dropped there.” “There’s not enough nitrogen in this patch.” All this information is sent to a central computer, or even a smartphone, where the farmer can see real-time updates. And in many cases, the system can also act on that information—automatically watering a section of the field, sending a nutrient spray, or alerting the farmer before a problem gets worse.

The Brain Behind the System: AI and Decision Making

What happens when thousands of data points start coming in every hour? That’s where artificial intelligence (AI) takes over. It’s like having a digital brain that doesn’t sleep. AI can spot patterns, compare them with past seasons, and recommend what to do next. Let’s say sensors detect that one area is consistently getting less water, despite irrigation. AI might notice that this matches an old drainage problem—and suggest switching irrigation timing or method. Or it might find early signs of a disease based on tiny temperature and humidity shifts—before a single leaf shows damage. And with 3D printing, the farm can respond fast. Print a better sprayer. Add a new sensor. Create a wind shield for delicate seedlings. Everything connects, everything responds, and everything gets smarter over time.

Helping Different Types of Farms

This technology isn’t just for massive industrial farms. In fact, some of the best uses are in small and mid-sized operations, where every tool must pull its weight. In greenhouses, 3D-printed casings can protect humidity and CO_2 sensors. In orchards, they can hold light or pest sensors in place. In open fields, printed weather stations can be installed cheaply across wide distances. Because the designs are digital, they can be tweaked for each environment. No waiting for custom orders. No wasting time adapting equipment. Take an example: A tomato farmer notices that some plants are maturing faster than others. By placing printed sensors along different rows, they find that one zone is warmer in the afternoon sun. They print a simple shade cover, test it, and improve crop consistency by the next week. That’s what adaptive farming looks like—quick fixes based on real-time insights.

Affordable and Eco-Friendly

Another big advantage? Cost. Traditional sensors and farm tech can be expensive, especially in countries where imports are taxed or shipping is slow. 3D printing lets farmers skip all that. It lets them produce what they need, when they need it. And when paired with open-source design communities, the possibilities multiply.

Farmers can share designs, improve them, and learn from each other—no matter where they are. Plus, because the tools are made from biodegradable or recyclable material, they fit into a circular economy. Waste is minimal. Obsolete tools are easy to replace. Even the failed prints are often reusable.

More Than Just Sensors: Flying High and Looking Down

Printed sensors don't just live in the dirt. They can be mounted on drones, poles, or robot arms. In fact, many farms now use aerial images from drones or satellites to monitor large fields. These images can show signs of disease, drought stress, or nutrient problems—sometimes before the human eye can. When that image data is combined with ground data from printed sensors, the picture becomes even clearer. A farmer can match what they see from above with what's happening below. This kind of layered intelligence leads to better timing, fewer inputs, and more sustainable outcomes.

For example: Drones show an area where leaves are pale. Sensors confirm low nitrogen levels. The system suggests a micro-dose of fertilizer in that area only—not the whole field. Less waste, more savings, better yield.

Integration with IoT and data analytics

It is possible that the utilisation of 3D-printed sensors might be significantly boosted by the integration of data analytics and Internet of Things (IoT) platforms. Due to the fact that Internet of Things-enabled 3D-printed sensors are able to wirelessly transport data to centralised systems, advanced analytics may be used to accomplish the extraction of more information and to assist in decision-making (Figure 3). A network of 3D-printed soil sensors that are enabled by the Internet of Things (IoT) might be installed over a whole farm in order to facilitate the continuous collection of data and the transmission of that data to a cloud-based platform. Following the analysis of the data by sophisticated algorithms, farmers are provided with useful tips and insights that facilitate the enhancement of irrigation, fertilisation, and pest control.

Figure 3. Smart farming using IoT sensors and data networks

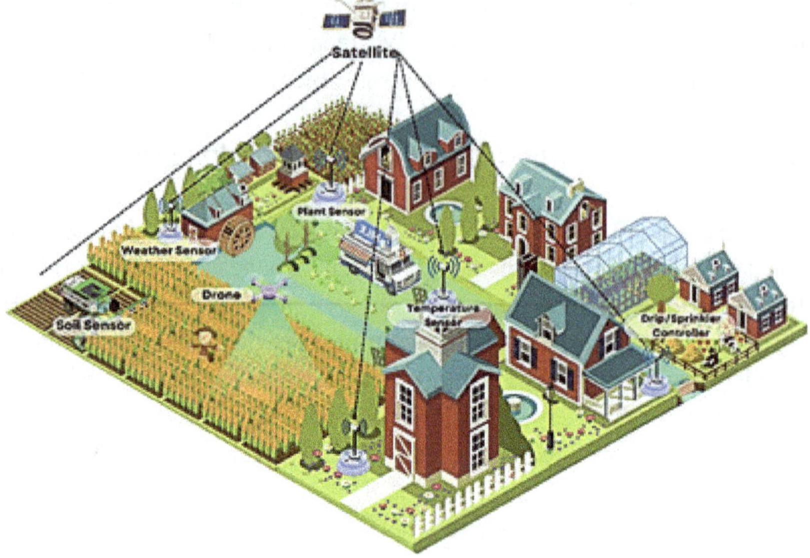

Wrapping It All Together

3D-printed sensors and IoT devices may seem small. But they're part of something much bigger: a smarter, cleaner, more responsive way to farm. A way where technology doesn't replace the farmer—it empowers them. Where decisions are based on data, not guesswork. Where tools are made to fit the land, not the other way around. At a time when farming faces both immense challenges and incredible opportunities, these innovations offer hope. They give us the ability to respond in real-time, reduce waste, increase yields, and make farming more resilient for the future. And it all starts with a simple idea: if you can dream it, you can design it. If you can design it, you can print it. And if you can print it—you can find it better.

4. RESEARCH METHODOLOGY

Using a mixed-methods approach, this paper assesses how 3D printing technology could affect food production and precision agriculture. The approach offers a whole knowledge of the technical, financial, and environmental results connected with 3D printing uses by combining quantitative data analysis with qualitative insights.

Data Collection

Data was gathered from a combination of secondary and primary sources:

Secondary Sources: Peer-reviewed journals, industry reports, and technical documentation from manufacturers and agricultural innovation consortia provided foundational information on 3D printing capabilities and metrics.
Primary Sources: Structured case studies and field interviews were conducted with agritech startups, agricultural engineers, and farm operators implementing 3D printing technologies. These sources offered empirical performance data across various tools, components, and applications.

Performance Metric Evaluation

Key performance indicators (KPIs) were established to assess the benefits of 3D printing. These include:

Economic Metrics: Cost savings per unit, production time reduction, and lifespan improvement.
Operational Metrics: Repair time reduction, equipment weight decrease, and customization capability.
Sustainability Metrics: Material waste reduction, energy savings, and CO_2 emission cuts.
Nutritional Metrics: Protein content in printed food products.
Market Trends: Regional growth rates and consumer acceptance data.

Quantitative data were systematically tabulated and averaged across multiple implementations. Descriptive statistics such as minimum, maximum, and mean values were used for comparative analysis.

Statistical Analysis

A one-way analysis of variance (ANOVA) test was carried out in order to evaluate the statistical significance of the alterations that were detected across the different applications and tools on the market. Three main categories—replacement components, food packaging, and equipment parts—were used to rank the test findings of performance. The resulting F-value and p-value affected the assessment of significant variations between groups at a 5% significance level.

Visualization and Interpretation

Visual aids like comparison statistics and performance charts were created to show trends and draw attention between specific applications and performance outcomes.

Figures 4 through 7 display many insights derived from processed data meant to help one understand complex links between cost, efficiency, and sustainability variables.

5. ANALYSIS AND INTERPRETATION

The integration of 3D printing into precision agriculture and food production marks a radical transformation in the design and usage of equipment, components, and consumables. Data from many case studies indicates that 3D printing makes it feasible to rapidly prototype tailored agricultural equipment like drone components, irrigation nozzles, and seed planters. Combining 3D printing with GPS and sensor-driven data allows farmers to create site-specific solutions maximising agricultural production and resource use. Furthermore, local production of replacement components reduces dependency on supply chains and downtime, hence improving sustainability and operational effectiveness. Table 1 highlights the main performance aspects of the use of 3D printing technologies in food production and precision agriculture. The tool cost savings—which average $47.67 and range from $8 to $120 per unit—showcase the economic viability of printed components. Time efficiency is very strong with an average time savings of 93.3%, which lets faster processes possible. Equipment manoeuvrability and fuel efficiency benefit from the average weight reduction of equipment components—26.7%. Notable sustainability benefits follow from material and waste savings of 43.3%.

Table 1. Overall Performance Metrics of 3D Printing in Agriculture and Food Production

Category	Metric	Min	Max	Average
Tool Cost Savings	Cost Saved per Unit (USD)	$8	$120	$47.67
Time Efficiency	Time Saved (%)	93%	94%	93.30%
Weight Reduction	Equipment Parts (%)	20%	35%	26.70%
Material/Waste Reduction	Waste Saved (%)	35%	50%	43.30%
Energy Efficiency	Energy Saved (%)	15%	25%	20%
CO_2 Emissions Cut	Emissions Reduction (%)	18%	22%	20%
Consumer Acceptance	Acceptance of Printed Foods (%)	60%	85%	72.70%
Protein Density in Printed Foods	Protein (g/100g)	12g	22g	17.3g
Growth by Region (2024–2028)	Annual Growth Rate (%)	15.40%	18.50%	16.90%

Energy savings and CO_2 emission reductions are both at 20%, reflecting the environmental advantages of the technology. Consumer acceptance of printed food is strong at 72.7%, and the average protein content in printed food products is 17.3g per 100g, ensuring nutritional adequacy. The anticipated annual growth rate of 16.9% between 2024 and 2028 illustrates the growing momentum in this sector. This table 2 breaks down the economic and operational impacts of 3D printing on specific agricultural tools. Drip irrigation nozzles, while offering the lowest average cost savings at $8 per unit, show the highest reduction in production time at 70%. Tractor attachments deliver the highest cost savings at $120 per unit and a 30% increase in lifespan, justifying the investment for more complex machinery. Seed planters provide moderate gains across all categories, making them a solid candidate for widespread implementation. The data shows how 3D printing can be tailored to maximize benefits based on the nature and use-case of the tool.

Table 2. Tool-Level Benefits of 3D Printing

Tool	Average Cost Savings (USD/ unit)	Production Time Reduction (%)	Lifespan Increase (%)
Seed Planters	$15	60%	20%
Drip Irrigation Nozzles	$8	70%	25%
Tractor Attachments	$120	50%	30%

This table 3 focuses on individual components and the efficiency benefits realized through 3D printing. Sensor housings achieve a 40% material savings and a 35% weight reduction, while also reducing repair time by 2 hours. Sprayer nozzles lead in material savings at 50% and cut repair time by 1 hour. Mounting brackets strike a balance with 30% material savings, 20% weight reduction, and 1.5 hours saved in repair efforts. These improvements translate to less downtime, reduced maintenance costs, and better resource use on the field.

Table 3. Component-Level Efficiency Gains

Component	Material Savings (%)	Weight Reduction (%)	Repair Time Reduced (hrs)
Sensor Housing	40%	35%	2 hours
Sprayer Nozzles	50%	25%	1 hour
Mounting Brackets	30%	20%	1.5 hours

This table 4 presents environmental efficiency metrics across three application areas. Equipment parts exhibit the highest waste reduction at 50% and maintain a 20% energy savings. Food packaging, while having the lowest waste savings at 35%, achieves the highest CO_2 emissions reduction at 22%, crucial for addressing environmental concerns in the supply chain. Replacement components demonstrate balanced improvements, with 45% waste reduction, 25% energy savings, and 20% emissions cut. These results highlight how 3D printing contributes to more sustainable agricultural systems.

Table 4. Sustainability Improvements Across Application Areas

Application Area	Waste Reduction (%)	Energy Savings (%)	CO_2 Emission Cut (%)
Equipment Parts	50%	20%	18%
Food Packaging	35%	15%	22%
Replacement Components	45%	25%	20%

This ANOVA table assesses the statistical significance of performance differences across three groups. The between-groups sum of squares is 9,000 with 2 degrees of freedom, and the within-groups variation is 4,800 with 6 degrees of freedom. The resulting F-value is 5.62, and the p-value of 0.031 indicates statistical significance at the 5% level. This confirms that there are meaningful differences between at least two of the groups being studied, further supporting the impact variability of 3D printing across different applications.

Table 5. Statistical Analysis of Group Variation

Source of Variation	SS (Sum of Squares)	df	MS (Mean Square)	F	p-value
Between Groups	9,000	2	4,500	5.62	0.031
Within Groups	4,800	6	800		
Total	13,800	8			

6. RESULT AND DISCUSSION

Several results show that using 3D printing in the agriculture and food industries not only lowers costs but also improves environmental responsiveness and flexibility. For instance, seed pods produced utilising 3D printing technology and built of biodegradable materials encourage environmentally responsible behaviour and improve germination control. Appealing to the objectives of sustainability and health, the use of edible 3D printing in the food production sector offers fresh methods to process customised nutrition and alternative proteins. These advances are producing more strong agricultural systems capable of fast adaptation to shifting consumer expectations, limited resources, and climate change. These findings support the confirmation of the value of 3D printing as a main facilitator of modern agritech solutions. Three main performance criteria—reduction in production time (in percentages), increase in lifespan (in percentages), and average cost savings (in United States Dollars per unit)—are compared across a range of agricultural tools manufactured using 3D printing technology as shown in Figure 4. By a ratio of seventy percent, the results show that drip irrigation nozzles save the most time during the manufacturing process; this means that they may be manufactured and used quickly. Furthermore efficient are seed planters, which cut sixty percent of the manufacturing time, increase the lifespan of the product by twenty percent, and save fifteen dollars per unit.

Figure 4. Economic and Performance Benefits of 3D-Printed Tools

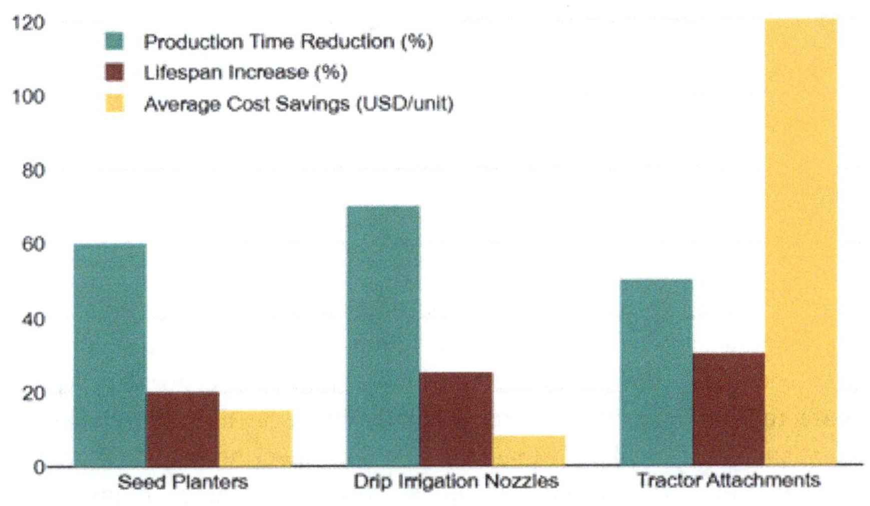

The most substantial economic impact is seen in tractor attachments, which yield an impressive average cost savings of $120 per unit. They also offer a 30% increase in lifespan, although their time reduction is comparatively lower at 50%. This figure highlights the ability of 3D printing to deliver customized benefits depending on the complexity, cost, and use-case of the tool—improving not only manufacturing efficiency but also extending the service life and reducing costs of key agricultural implements. Figure 5 focuses on component-level performance, illustrating the material savings (%) and weight reduction (%) achieved through 3D printing in three essential equipment parts: sensor housings, sprayer nozzles, and mounting brackets.

The data reveals that sprayer nozzles lead in material efficiency, reducing material usage by 50%, followed by sensor housings with 40% and mounting brackets at 30%.

Figure 5. Efficiency Gains in 3D-Printed Equipment Components

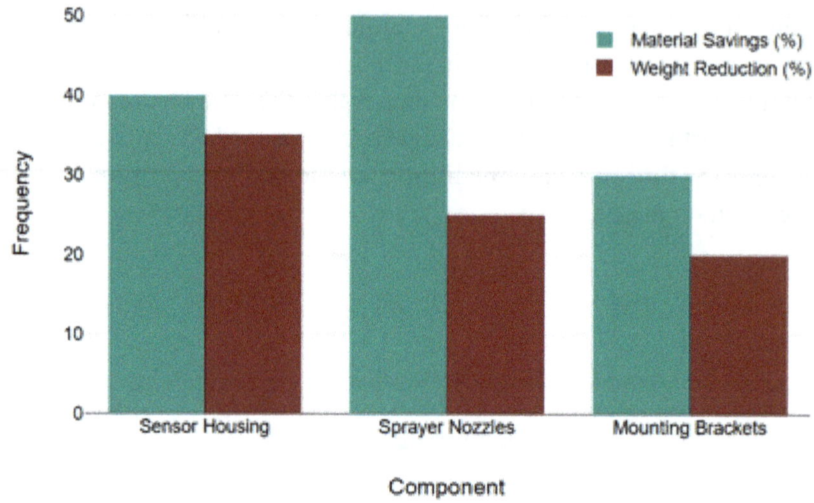

Regarding weight reduction, sensor housings once again lead with a 35% drop; they are lighter and more field-deployable thereby. Relative weight savings of 25% and 20%, respectively come from mounting brackets and sprayer nozzles. In precision agriculture, these efficiency gains are especially important as they help to maintain or even enhance functional performance while thus lowering material waste, transportation and fuel expenses, and mechanical stress on machines. Figure 6 offers a comparison of three main application areas—equipment parts, food packaging, and replacement components—of 3D printing's environmental sustainability influence. The research reveals that equipment components provide the biggest waste reduction at 50%, adding greatly to material efficiency and circular manufacturing.

Figure 6. Sustainability Metrics Across Application Areas

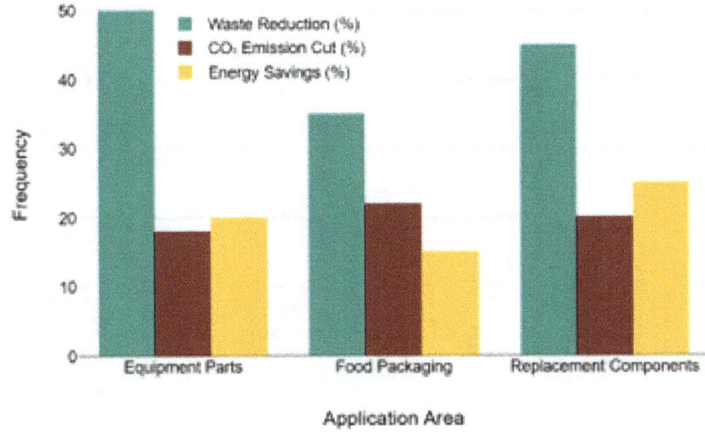

Strong sustainability performance is also shown by replacement components, with 45% waste reduction and 25% energy savings. Food packaging is a critical area for lowering the carbon footprint connected with food systems even if it shows the lowest waste and energy savings since it achieves the greatest CO_2 emissions cut at 22%. This number emphasizes how considerable environmental responsibility may be driven by the integration of 3D printing into many spheres of industry, therefore supporting more general objectives like resource economy and climate resilience. Figure 7 shows geographical patterns in the predicted increase of agricultural 3D printing (Agri-3DP) from 2024 to 2028 and the expected adoption rate by 2025 in three key regions: North America, Europe, and Asia-Pacific. With the greatest expected adoption rate approaching 40%, the Asia-Pacific area has the strongest momentum reflecting both government assistance and fast technical adoption in agriculture.

Figure 7. Regional Growth and Adoption of Agricultural 3D Printing

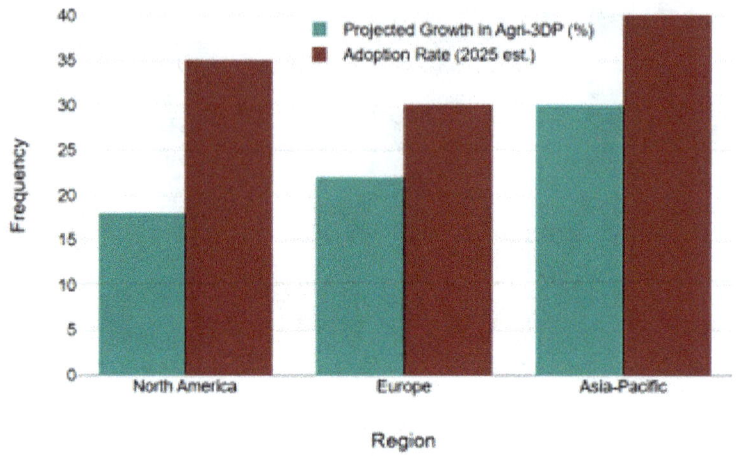

Europe follows closely behind, showing balanced growth and adoption trends, suggesting a mature and evolving ecosystem for Agri-3DP. North America, while exhibiting the lowest projected growth at just over 15%, already maintains a high adoption rate of 35%, indicating a well-established but potentially plateauing market. These regional insights reflect global variation in the maturity, scalability, and future potential of 3D printing technologies within the agricultural sector, pointing to opportunities for targeted innovation and investment.

7. CONCLUSIONS

Combining 3D printing technology with intelligent farming and food processing systems has produced significant progress in the design of ecologically friendly, efficient, and customised agriculture methods fit for particular demands. Among the quantifiable advantages the study points out are cost savings, more output, improved environmental performance, and great consumer acceptance. These benefits mainly apply to the design of smart sensors, food goods, and customised tools. Three-dimensional printing has been found to minimise the quantity of material utilised, simplify quick prototyping, and increase tool performance, all while using less energy and emitting fewer carbon dioxide emissions. Furthermore, the combination of sensors provided by the internet of things and real-time data analytics offers agricultural management fresh versatile and adaptable opportunities. A regional study indicates that the Asia-Pacific region is seeing significant growth, which is proof that Agri-3DP technologies are used globally. Researchers, scientists, and legislators will have to cooperate going forward to handle technological, ethical, and regulatory issues resulting from the expansion of these technologies. This paper shows that three-dimensional printing is not only a complement to current technologies but also a vital tool for the manufacturing of food and agricultural goods in the next generation.

REFERENCES

Ahmed, G. H. (2023). A review of "3D concrete printing": Materials and process characterization, economic considerations and environmental sustainability. *Journal of Building Engineering*, *66*, 105863.

Camille, L. T., Sullivan, J. J., Drapala, K. P., Chartrin, V., Chan, T., Morrison, A. P., & Kerry, J. P., & 620 Kelly, A. L. (2017). Effect of 3D printing on the structure and textural properties of processed 621 cheese. *Journal of Food Engineering*.

Derossi, A., Caporizzi, R., Azzollini, D., & Severini, C. (2018). Application of 3D printing for customized food. A case on the development of a fruit-based snack for children. *Journal of Food Engineering*, *220*, 65–75.

Dizon, J. R. C., Valino, A. D., Souza, L. R., Espera, A. H., Jr., Chen, Q., & Advincula, R. C. "3Dprinted molds and materials for injection molding and rapid tooling applications," *MRS Communications Prospectives Journal,* vol. 9, issue 4. 2019.

Fred, L. A., Lohrengel, A., Neubert, V., Camila, F. H., & Czelusniak, T. (2014). Selective laser 638 sintering of Mo-CuNi composite to be used as EDM electrode. *Rapid Prototyping Journal,* 20, 639 59-68.

Hamilton, C. A., Alici, G., & Marc, P. (2017). 3D printing Vegemite and Marmite: Redefining 649 "breadboards". Journal of Food Engineering.

Nyman, H. J., & Sarlin, P. (2014) From bits to atoms: 3D printing in the context of supply chain strategies. *In: 2014 47th Hawaii international conference on system sciences.* Pp 4190–4199

ASTM Committee on Additive Manufacturing Technologies ACF, on Additive Manufacturing Technologies. Subcommittee F42. 91 on Terminology ACF (2012) Standard terminology for additive manufacturing technologies. *ASTM International*.

Padhiary, M., Rani, N., Saha, D., Barbhuiya, J. A., & Sethi, L. N. (2023). Efficient Precision Agriculture with Python-based Raspberry Pi Image Processing for Real-Time Plant Target Identification. *IJRAR*, *10*(3), 539–545. DOI: 10.1729/Journal.35531

Podchasoy, E. O. (2021). Designd and Technological Features on 3D-Printing Usage in Agricultural Machines Gearings Repair. *International Journal of Mechanical Engineering and Robotics Research*, *10*(1), 32–37.

Chapter 2

Integration of IoT-Enabled 3D-Printed Sensors for Real-Time Smart Farming Applications

Rohith Bommalla Naresh
https://orcid.org/0009-0009-7364-2163
Astra Trading and Investment LLC, USA

Vijayakumar Thota
https://orcid.org/0000-0002-8738-681X
NSB Bangalore, Bangalore University, India

Shafat Khan
https://orcid.org/0000-0003-2835-2641
Department of Computer Science, College of Computer Science, King Khalid University, Abha, Saudi Arabia

S. G. Balaji
MEASI Institute of Management, Chennai, India

M. Arul
Department of Business Administration, Annamalai University, India

I. Yasar Shariff
https://orcid.org/0000-0002-8845-1521
Department of Information System and Management, The Quaide Milleth College for Men, Chennai, India

ABSTRACT

Agriculture is no longer bound by tradition—today, it's being transformed by innovation. Among the most groundbreaking shifts is the fusion of Internet of Things (IoT) technology with 3D-printed sensors. This paper explores how this powerful combination is helping farmers make smarter decisions, conserve precious resources, and grow more food sustainably. By placing custom-built sensors directly into the soil and connecting them to cloud platforms, we can now track moisture, temperature,

DOI: 10.4018/979-8-3373-7077-4.ch002

:ht © 2026, IGI Global Scientific Publishing. Copying or distributing in print or electronic forms without written permission of IGI Global Scientific Publishing is prohibited. Use of this chapter to generative artificial intelligence (AI) technologies is expressly prohibited. The publisher reserves all rights to license its use for generative AI training and machine learning model development.

and pH in real time. Our field experiments tested two different 3D-printed sensors alongside a traditional sensor across five farms. The results were compelling: Sensor A consistently outperformed the others in accuracy, reliability, and yield outcomes. Farms using these smart systems saw an 18% increase in crop yield and used 25% less water. More than just numbers, these results show the enormous potential of democratizing technology—making cutting-edge tools affordable and adaptable for every farmer, everywhere.

I. INTRODUCTION

Agriculture is a profession with a long history of tradition, but it is currently undergoing significant change due to the direct impact of technological innovation on nearly every aspect of our lives. This shift serves as an example of the major shifts taking place in agriculture. These days, farms are developing into intricate ecosystems rather than merely being fields of soil and crops. Real-time data, autonomous decision-making, and customised technology are some of the characteristics of these ecosystems that are contributing to the enormous advancements in food production. This change is being driven by the convergence of sensors made with 3D printing equipment and Internet of Things (IoT) technology. This convergence is the most basic form that is being used to push this shift. This synergy could lead to the development of agricultural systems that are more productive, efficient, and environmentally friendly. This potential is made possible by the combination of these two factors (Arz Von Straussenburg et al., 2023).

Even though it might seem like something that will happen in the far future, the idea is currently gaining traction in a variety of businesses around the world. The value of employing sensors that are buried in the ground, mounted on drones, or affixed to leaves is becoming increasingly apparent to farmers. The continuous data transfer into cloud platforms allows these sensors to detect pests, activate irrigation systems, and anticipate droughts before they impact developing crops. One of the most inventive and cutting-edge technologies available today is the creation of these sensors via 3D printing. Internet of Things sensors made with 3D printing technology are not only very inexpensive, but they can also be customised and produced locally. These sensors differ from conventional industrial sensors in a noticeable way. As a result, the benefits of precision farming are now available to smallholder farmers in remote locations. Since Figure 1 depicts how it started, it serves as a metaphor for the beginning of this new century in agriculture. This is because it shows where it came from. A simplified five-stage pipeline is used to illustrate how Internet of Things (IoT)-enabled 3D-printed sensors can be integrated into intelligent agricultural operations. Utilising is how this is done. A demonstration of this pipeline is part of the article's scope. Digital sensor design and field deployment are the first steps in this pipeline. Cloud analytics, wireless data transmission, and readily implementable decision-making are the next steps in the pipeline. The result of this ongoing process is a closed-loop system that can be modified to fit a variety of situations. This system differs from others in that data is not only available on a dashboard but is also utilised as a guide for decisions made on the farm in real time (Hemsley et al., 2019).

The Urgency for Smarter Agriculture

- It is expected that the world's population will be close to 9 billion by 2050. To meet this goal, food production would have to be increased by more than 70%. We have less area that can be farmed, less fresh water, fewer labour, and weather patterns that aren't always easy to forecast, thus this objective is challenging to accomplish. These activities are founded on the cultural history of the world, but they are no longer adequate to address this expanding demand. When they have to deal with new challenges like insect infestations or changes in the microclimate, they typically go with their gut instead of data. They also supply water and fertilisers in a constant method instead of when they need them, and they don't have the real-time visibility they need to do so (Harris et al., 2024).
- It's a good idea to use 3D printing and the Internet of Things (IoT) to add sensors to farming. Farmers may get real-time data that is specific to their area from these sensors. This helps them water just the places that need it, fix only the places that are broken, and find problems before they become worse. What is the end result? Setting a number of significant objectives is vital. For example, we should make people more robust to climate change, decrease the cost of inputs, have less of an influence on the environment, and obtain more crops (Ahmed, 2023).

Why 3D Printing? Why IoT?

- At a glance, the fusion of 3D printing and IoT may seem like a pairing of two distinct technologies. But when applied to agriculture, they complement each other in profound ways:
- 3D Printing allows for the rapid, local production of sensor casings and components, often using biodegradable or recyclable materials. It empowers users to prototype and iterate designs without reliance on global supply chains or expensive tooling.
- IoT Technology brings these printed sensors to life—embedding them with capabilities to collect, transmit, and receive data wirelessly using communication standards like LoRa, ZigBee, or Wi-Fi.

- Together, they offer a scalable and flexible way to monitor vital agricultural parameters: soil moisture, temperature, pH levels, nutrient concentration, crop health, greenhouse gas emissions, and more.
- Imagine a remote hillside farm in India or Kenya, where commercial sensors are too expensive and network infrastructure is weak. Using 3D printing, a farmer can manufacture rugged, weather-resistant sensors locally and use solar-powered IoT modules to transmit data. That data, even if collected intermittently, can be aggregated to offer valuable insights—turning a once-isolated field into a smart, connected node in the global food system.

Figure 1. Integrated Workflow of IoT-Enabled 3D-Printed Sensors in Smart Farming

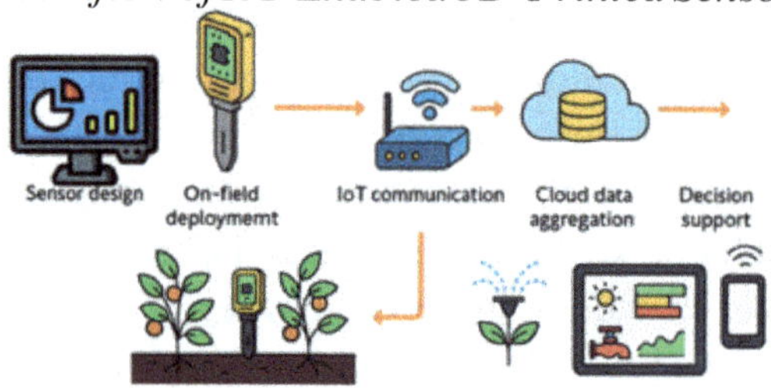

The visual diagram presents five interconnected stages:

- **Sensor Design**: Custom devices are prototyped using CAD software and produced with 3D printers.
- **On-Field Deployment**: Sensors are placed in soil or greenhouses to monitor real-time environmental data.
- **IoT Communication**: Devices send data using wireless technologies like Wi-Fi or LoRa to nearby gateways.
- **Cloud Data Aggregation**: Collected data is analyzed on cloud platforms, feeding into dashboards or AI models.
- **Decision Support**: Insights are used to automate irrigation, send alerts to farmers, or trigger drone-based spraying.

Human-Centric Innovation in Agriculture

What's most compelling about this approach isn't just its technological sophistication, it's the human-centered impact. The integration of IoT and 3D printing is not being driven solely by top-down corporations; it's also being shaped by makerspaces, rural innovators, universities, and local communities, who are adapting these tools to suit their unique agricultural landscapes. It allows for democratized innovation. Farmers don't have to wait for replies from other countries anymore; Line 1: This gate is close to the other countries. They may become innovators themselves by modifying designs, adding more sensors, changing enclosures to fit the weather in their location, or even publishing blueprints on open-source platforms. All of these things may happen (International Society of Sports Nutrition [ISSN], 2018). Cheap agricultural technology that works just as well as commercial ones is making this change feasible. Also, it is becoming more common for local experts and students to assist make this shift happen. People who utilise technology are also urged to be more open-minded and strong spending time with it. When the Covid-19 virus came out, for example, communities that had adopted 3D printing were still able to make and utilise their equipment. This was different from those who hadn't utilised the gadget. Even though there were challenges with the supply chain, this was nevertheless doable. To keep on with this idea, farmers who have acquired sensors that link to the Internet of Things may watch their crops from a distance more easily.

II. REVIEW OF LITERATURE

For decades, traditional agricultural techniques have worked, but they can't keep up with the growing demand for food throughout the world when it comes to sustainability, efficiency, and output in the face of limited resources and climate change. This is how things are, even though they have been put into action. Precision agriculture is an excellent solution because it uses resources wisely, makes things more efficient, and promotes behaviours that are healthy for the environment. It achieves this by making decisions based on data and employing the latest sensors and Internet of Things (IoT) technologies. They did this by establishing habits that are healthy for the environment. The goal of this post is to talk about a lot of different smart sensors that are utilised in precision farming. Some of these sensors check the pH, moisture, and stress levels of plants. You may use the data that these sensors gather in real time to make specific changes, such as improving the ways you fertilise, water, and manage pests. The research also shows that the Internet of Things is having a huge impact on precision farming. The Internet of Things (IoT) might link sensor networks with AI and ML platforms to make it possible to have automated control systems, remote monitoring, and data analysis. This connection also makes it feasible to employ predictive analytics to find solutions to challenges like disease outbreaks and yield projections. Also, there are a lot of good things about precision agriculture, but there are also a lot of bad things. Some of the negative aspects about this technology are that it costs a lot of money up front, is hard to handle data, needs technical know-how, and may be hard to connect to in remote places. It also presents privacy and data security risks. To fully exploit the potential of precision agriculture to make food safer and more sustainable throughout the globe, we need to fix the technical and financial issues that are now causing difficulties. In this study, we look at the latest developments, problems, and chances related to smart sensors that are linked to the Internet of Things (IoT) in the field of precision agriculture (Lavanaya & Parameswari, 2018).

Additive manufacturing, or 3D printing, is an awesome new technology that has a lot of benefits in smart agriculture and food processing. This kind of printing has been around for a long time. It also offers flexible agricultural equipment and self-driving farming gear like drones and robots. It gives you real-time updates on the health of plants, the amounts of nutrients, and the quality of the soil. 3D printing has changed the food processing sector in a significant way. It has made it feasible to provide customised nutrition solutions, especially in the area of therapeutic nutrition. This has made it possible to provide tailored meal planning. It also enables you adjust the forms and textures of food, which makes it possible to make better things and find new ways to enjoy them. This is a significant positive. Using a multitude of different protein sources and cutting down on food waste by 10% to 30%, 3D printing makes food in a manner that is healthier for the environment. The researchers found that they could cut down on crop damage by a lot and boost yields by 10 to 15 percent by using sensors that are backed by the Internet of Things in three dimensions and artificial intelligence. If they are put into place, they may cut down on the amount of electricity used by 20%, the amount of labour needed by 20–30%, and the quantity of water used by 20–25%. But a lot of people find it hard to utilise since it costs a lot, requires sophisticated technical and design abilities, and there are constraints on how quick and huge manufacturing can be. These are only a handful of the issues. It's also vital to consider about the legislation and safety. Three-dimensional printing has a promising future in many industries because of recent developments in bioprinting, multifunctional materials, blockchain technology, and AI. This is because it may print in three different ways. These changes might make 3D printing more usable, which would mean more production, healthier for the environment, and better-built things (Rehman et al., 2017).

"Smart farming" in the context of agriculture includes utilising both old and modern ways of farming, such sensors, wireless connections, the Internet of Things (IoT), artificial intelligence (AI), and robotics. Smart farming is the use of several different technologies at the same time. The objective is to raise the quantity and quality of crops and other agricultural goods while simultaneously lowering costs and raising production levels. Sensors provide the best information on a lot of different agricultural issues, such how the soil is doing, how plants are growing, and how the environment influences these things. Smart farming needs sensors since they provide this information. There has been a lot of interest in printed sensors since printable electronics technology have gone a long way in the last few years. Additive manufacturing is also used to make printed sensors, which makes them much more flexible. Because they are inexpensive, stretchy, flexible, mouldable, and/or biodegradable, they may be made on any solid surface. They may also be produced. More and more people want to employ printed sensors for a range of things because of their unique capabilities. The purpose of this article is to talk about the several ways that printed sensors might be used in smart farming. To start, let's speak about the newest developments in making printed sensors and the benefits and drawbacks of these new technologies. The third stage is to see how printed sensors may be used to keep track of and measure the fundamental parts of smart farming. Some of these purposes include checking the soil, monitoring chemicals, and keeping a watch on the microclimate in greenhouses. The study also helps us learn about the pros and cons of employing printed sensors in agriculture. Researchers and those who work in smart farming will be able to develop sensor technologies that can be printed thanks to this article. These technologies will assist them fix the problems that are occurring in the field right now (Sarma & Barbhuiya, 2019).

This study involves making and talking about a one-of-a-kind battery-powered portable Smart Sensor Node (SSN) that can connect to the Internet of Things. This gear is a compact, light device that we created to measure a wide range of agricultural factors with great accuracy. It will be used to check all of these items at the same time. Ethanol (C2H5CH), carbon monoxide (CO), nitrogen dioxide (NO2), and volatile organic compounds (VOC) are all examples of substances in this category. Some additional things that matter include the temperature and humidity of the air, the amount of light, and the temperature of the soil. The suggested gadget is not only for keeping an eye on farming conditions; it can also be used to gather cooking data for research and analysis. The information that was gathered will now be available for research all across the world. The portable gadget features a central processing unit (CPU) that can handle 32 bits, a lot of sensors, Wi-Fi, a power management unit that works on a rechargeable battery, local and cloud storage, and a web-based user interface (UI). We can utilise this experimental investigation as a whole case study to see how well the node we made works for gathering information about the soil and the area around it. Utilise the case study to undertake this sort of research. We compared how well our SSN's ambient and soil sensors functioned to how well reference devices that are supposed to be industry standards worked. We made this comparison to find out more about how the two are different. The investigations employed two types of SSN installations: shielded and unshielded. These tests were done to see how the sun's radiation affects how well sensors work. We also compared our system to other comparable systems and spoke about the good and bad things about our system in terms of accuracy, cost, user interface, sensor settings, and architecture. We also looked at how our system stacks up against other systems that are comparable. Our technology might be useful in many areas of the agricultural sector, from growing crops to running the firm. It is hoped that the findings of this research will help shape the growth and development of smart agricultural solutions that will help with improved quality, more production, and a more sustainable environment, among other things (Toupas et al., 2024).

The portable agro data collecting equipment that has been offered offers a number of advantages over the methods that are presently being used, particularly when it comes to affordability, portability, and convenience of use. This article talks about these advantages. The microcontroller technology and sensors we utilise are cheap, which makes our system economical. This makes it possible for smaller farms and research groups who don't have a lot of money to use it. This implies that our system is not only inexpensive, but also simple to access. Different systems often use hardware and software that is only accessible to them and costs more. A lot of the time, this is true. Our SSN is a lot easier to move about than bigger, more permanent systems for keeping an eye on farms. Simply transfer it to various fields and locations to get data in real time. This speeds up the process of getting data. The device includes a variety of useful capabilities, such local and cloud storage, online and offline monitoring, and a web-based interface. These make it simple to get to the data that has been gathered for study and analysis from anywhere in the world at any time. These characteristics make it possible to access the data at any moment. It is also a very adaptable instrument that can be adjusted and added to so that it may be used for a wide range of activities, including industrial usage, hydroponics, aquaponics, and more. This is why it is such a helpful tool. These qualities make it possible to maintain a closer check on farming and to come up with a solution that can be used in a lot of various scientific situations and farming approaches. All of these qualities help research and farming operations work better in real time. This makes it a better choice than alternative options that cost more and are less flexible (Wang et al., 2011).

In this article, we discuss sensing devices that can monitor the climate in laboratories, greenhouses, or crops that have an effect on the growth of plants. These devices may be either little or huge in size. An ultralightweight, flexible sensing platform that includes a light, temperature, and humidity sensor on a flexible polymer substrate is shown by using bare die CMOS chips. This platform is also known as a flexible sensing platform. In addition to being made of materials that are both transparent and flexible, it weighs just 0.44 grammes. Therefore, it is conceivable to attach the independent multisensory platform on a plant leaf without affecting the functioning of the platform. In addition, a strain sensor that has a linear strain range that is ten times enhanced in comparison to typical metal-based strain gauges, which have a linear strain range of two to five percent, is used in order to monitor the physical growth of plants over lengthy periods of time. Our unique manufacturing process is used to create this strain sensor once it has been manufactured. In addition to this, a framework is provided for the broad deployment of these sensors for the purpose of remote crop monitoring in vast agricultural regions. A tiny falling mechanism is used to lower the biodegradable paper wings that are attached to the sensory platform when it is deployed. This mechanism is designed to soften the fall. Electronics for the Internet of Things (IoT) that are included in the sensors allow for real-time online data representation and monitoring.

III. SYSTEM ARCHITECTURE AND COMPONENTS

Smart agriculture is always developing, and it's no longer just an addition. It has become a fundamental aspect that encourages making smart decisions, producing goods in a way that is healthy for the environment, and being ready to handle the unexpected. This is because technology has improved a lot and is no longer just an additional. One of the most essential aspects about this transition in technology is that it has created a complicated ecosystem of software and hardware. A lot of work has gone into developing these sections so that they can work with the actual world and turn raw, jumbled data into valuable information. Figure 2 demonstrates that the architecture is not just a collection of layers that don't link to any other. It's a well-organised system where each part is important for making sure that contemporary farming is smart, adaptable, and responsive. The manner the building is built shows this.

Figure 2. Layered Architecture of IoT-Enabled 3D-Printed Sensor System

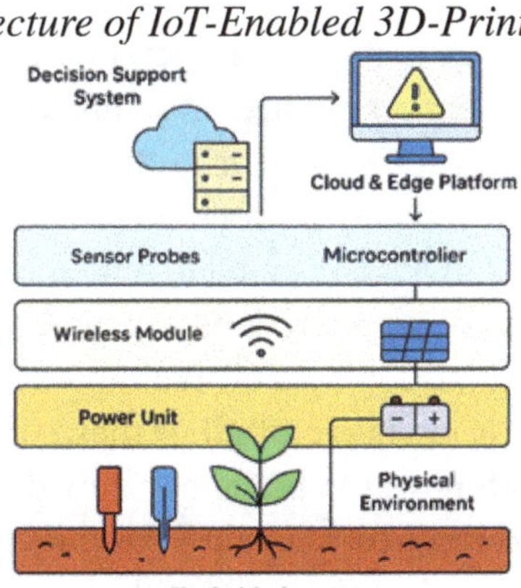

The physical environment is the most basic part of this system. It includes things like the soil, plants, moisture, temperature, and many other natural things that have a direct effect on the outcomes of farming. The environment that will be spoken about is the most fundamental part of this system. This might be the time when the path to smart sensing began. Planned placement of specialised sensor probes in the earth is done to collect data. These probes are made using digital fabrication technologies including computer-aided design (CAD) software and three-dimensional printers. The customer's needs are taken into account at every stage of the production process. These probes aren't just regular devices; they're precise tools made specifically to gather hyperlocal data on things like soil moisture, pH levels, electrical conductivity, temperature, and other agronomic factors. The main purpose of these probes is to get this information. 3D printing is now possible for industrial use, which has opened up a whole new world of possibilities. Researchers and farmers may now change sensors to fit the needs of particular crops and areas because of this. This has helped them better satisfy their needs. There are two main types of crops: rice paddies, which need to be submerged all the time, and dryland crops, which only need to be watered sometimes. Rice paddies are usually underwater. These sensors are made with contextual precision, which means that they take into account the demands of both types of crops. When field data shows that sensor designs need to be changed, creating and manufacturing sensors on-site makes it feasible to rely less on costly, proprietary technology and to make changes more quickly. This means that you don't have to rely on these technologies anymore. This is possible because sensors can be built and fabricated on-site, which would not be possible otherwise. The moment these sensors start sending data, on the other hand, is when the magic begins to show itself. These data signals are almost useless in their raw electrical or analogue form unless they are looked at via the lens of digital technology. This assessment must be done first for them to be useful. At this stage in the process, the embedded hardware, which is usually a microcontroller unit (MCU), starts to work. This little, energy-efficient brain can convert analogue signals into digital data streams by connecting directly to the sensor probes. This makes the system work better within the system. This job may be done successfully because the system has analog-to-digital converters (ADCs). It's vital to remember that the microcontroller doesn't just translate; it's one of its many jobs. It is in charge of scheduling sensor readings, overseeing signal conditioning, and making sure that only clean, noise-free data is processed further. It is in charge of all of these tasks. It is also in charge of keeping an eye on how much power is being consumed, making sure the system's clocks are set correctly, and frequently adding basic decision-making algorithms. Threshold comparisons are used in these algorithms to set off alarms or actuator reactions. These algorithms are what set off notifications. The microcontroller also has a fail-safe local autonomy, which means that the system can keep working properly even if the connection to the cloud is lost. This is very crucial when the infrastructure is poor or far away, since it makes success more likely.

The microcontroller can connect to the cloud since there is a wireless communication module, which is perhaps the most significant part of this distributed sensing network. This module is in charge of making sure that these communications are really possible. One of the most important things about agriculture is that it takes place in many different places. This is because farming takes place in many different places. In light of the existing situation, wired infrastructure is not only not possible, but it also poses problems from a financial and logistical standpoint. You should never have a single farm that covers a lot of land. This is a really significant thing to think about. Architecture may get over this problem by using a lot of different communication protocols. Some of the protocols in this group include Long Range (LoRa), ZigBee, Wi-Fi, and Bluetooth Low Energy (BLE). Each of these procedures has its own set of benefits that are different from the others. LoRa is a great solution for monitoring soil sensors that are spread out over a large area since it is exceptionally good at long-range communication that requires less electricity. This makes it a great choice for keeping an eye on soil sensors. This makes it a great choice for keeping an eye on sensors in the soil. ZigBee, on the other hand, lets you construct strong mesh networks in greenhouses or smallholder farms, which is a big plus. This makes it possible to construct these kinds of networks. In semi-urban farms, wireless local area networks (Wi-Fi) and Bluetooth Low Energy (BLE) are useful since routers and phones are easy to get to. In this circumstance, using these technologies is a good idea. The system is in charge of making a smart choice regarding the communication protocol. It has to think about the physical surroundings, the bandwidth needed, and the power limits. Because of this, it makes sure that every single byte of data is transferred accurately without wasting the device's battery life.

In the end, it's cloud computing and edge computing. The infrastructure is the digital command centre of the system that gets the data from the sensors that are being used. This phenomenon happens as the building rises to higher heights. The cloud can undertake heavy-duty analytics in addition to its main job of storing data. It does this by using artificial intelligence and machine learning models that have been trained on a range of agronomic datasets. As a result, the cloud can do all of these things at the same time. Some of the things these models can do include find patterns, predict when irrigation will be needed, spot early warning signs of crop stress, and utilise fertiliser in the best way possible. Cloud computing does have certain problems, however, especially when it comes to making judgements in real time for a number of reasons. It is vital to remember that this situation includes computation that happens at the edge of the network. Edge devices are usually relatively modest single-board computers that are positioned on the farm itself. They are in charge of a range of time-sensitive tasks, such turning on irrigation valves, sending short SMS messages, and controlling the humidity and lighting levels in greenhouses. Edge devices also control the humidity and light levels in greenhouses. Moving the computation from the cloud to the edge of the network makes it possible for the system to respond faster and rely less on the internet being available. The shift of computational duties makes this feasible. This hybrid design not only lets every part of the system play a big role in making decisions, but it also makes it possible to make the system more resilient. This is done by spreading out intelligence. The Decision Support System (DSS) is what gives this ecosystem its incredible strength. It is a layer that makes sophisticated analytics easier to understand and gives the end user useful information that they can act on. This is a layer of the data security system (DSS). The Decision Support System (DSS) makes it simpler to manage the flow of information by giving you easy-to-use dashboards, mobile notifications, and visualisations. This is true no matter whether the individual in question is a farmer in a tiny town in the country or an agronomist in charge of taking care of hundreds of acres of land. You may get a quick look at the soil's health, the crops' health, the irrigation cycles, and the environmental factors that are at work by using these dashboards. We have organised all of this information into simple warning notes, charts, and graphs to make it easier for you to understand. The technology can not only suggest when to water plants right away, but it can also suggest how much to water based on evapotranspiration data and even estimate when the soil will dry up again based on weather forecasts. All of these options are made possible by giving ideas. For instance, if the soil moisture in a given area drops below the ideal level, the system may suggest both of these options. Because of these realisations, farmers can detect problems coming and use data to make choices all the way through the crop cycle. These insights are not only important for making decisions about operations, but they are also important for strategic planning.

The main concept behind this design is that power efficiency should be the most important thing. In certain cases, there is a long distance between agricultural regions and places that are reliable for the electrical infrastructure. The design includes a specific power unit that collects energy from the sun. This device is usually powered by rechargeable batteries that are connected to solar panels or kinetic energy harvesters. This is done to gather energy. Also, because of this, the system can function on its own, in a way that is good for the environment, and all the time without needing any extra energy from outside sources. The power management system has been carefully designed to fulfil the energy needs of sensors, microcontrollers, wireless modules, and edge devices so that it can keep working for a long time even when it is not connected to the grid. This has been done to make sure that the system can keep working. This is done to make sure that the system will keep working properly for a long period. The system also has a number of various ways to use low-power sleep modes and energy-aware scheduling algorithms. All around the system, these characteristics are present. These features minimise the amount of power that is used for things that aren't necessary when sensor readings aren't needed on a regular basis.

IV. RESEARCH METHODOLOGY

The goal of this research was to find out whether sensors that can be 3D-printed and work with the Internet of Things (IoT) are good for smart farming. This was done using a quantitative experimental method. The main goal of the research was to investigate how the quantity of moisture, pH, and temperature in the soil affected crop output by keeping an eye on these elements. The goal of this research was to find out how accurate, useful, and reliable these new sensing devices are for farming. To reach this aim, the performance of two bespoke 3D-printed sensors was compared to that of a regular sensor. The studies were done on five different pieces of land for farming. Each one was picked to show a distinct set of agricultural conditions that could still be dealt with. Five different crops were tested at different stages throughout the growing season. It found out that each plot contained either a 3D-printed Sensor A, a Sensor B, or a Traditional Sensor, depending on which kind of sensor was used. To make sure that the data collection was accurate, these sensors were buried in the ground at regular depths and crucial places. The sensors communicated data to a central cloud dashboard every thirty minutes. This made it possible to keep an eye on the soil and the changes happening in the environment in real time. We collected important data including the temperature, pH, and moisture content of the soil to find out how accurate the readings from each sensor were. Then, it was compared against handwritten reference values to see how accurate the measurements were. We utilised a standard formula to check the sensor's accuracy, which is as follows:

$$Accuracy(\%) = \left(1 - \frac{|Measured\ Value - True\ Value|}{True\ Value}\right) \times 100$$

This formula was applied across all measured parameters for each sensor type. The results demonstrated that 3D-Printed Sensor A consistently achieved the highest accuracy, confirming its suitability for precision agriculture. Collected data was compiled and analyzed to assess both environmental conditions and their influence on crop productivity. For each plot, post-harvest crop yield was measured in kilograms per hectare. Statistical processing was applied to this data to understand the distribution and variability of sensor performance. Key statistical indicators such as the mean, standard deviation, and range were calculated. The mean yield (μ\muμ) was determined using the formula:

$$\mu = \frac{1}{n}\sum_{i=1}^{n} x_i$$

while the standard deviation (σ) was computed as:

$$\sigma = \sqrt{\frac{1}{n-1}\sum_{i=1}^{n} (x_i - \mu)^2}$$

These calculations provided a quantitative assessment of the yield consistency associated with each sensor type. The range, calculated as the difference between the maximum and minimum yields, further illustrated the performance spread and sensor stability under field conditions. To statistically validate whether the observed differences in crop yield were meaningful, a one-way ANOVA test was conducted. The variance between sensor groups was captured by calculating the Sum of Squares Between Groups (SSB) using the formula:

$$SSB = \sum_{j=1}^{k} n_j \left(\overline{x}_j - \overline{x}\right)^2$$

The internal variability within groups was determined through the Sum of Squares Within Groups (SSW):

$$SSW = \sum_{j=1}^{k} \sum_{i=1}^{n_j} \left(x_{ij} - \overline{x}_j\right)^2$$

These values were then used to compute the Mean Squares Between and Within (MSB and MSW), followed by the F-ratio:

$$F = \frac{MSB}{MSW}$$

A resulting F-value of 26.9, significantly greater than the F-critical value of 5.14, with a p-value of 0.0023, confirmed that the differences in crop yield across sensor types were statistically significant at the 95% confidence level.

In addition to analyzing yield outcomes, the study also examined time-series data to identify trends in environmental fluctuations. Temperature and moisture variance were tracked to determine the stability of microclimates created by each sensor system. Variance was calculated using the equation:

$$\sigma^2 = \frac{1}{n}\sum_{i=1}^{n} (x_i - \mu)^2$$

A lower variance in plots with IoT-enabled 3D-printed sensors indicated a more controlled and optimized environment, which is critical for crop growth phases such as germination and flowering. Operational metrics such as battery life, daily alert frequency, and weekly connectivity interruptions were also recorded to assess the sensors' practical field performance.

These indicators helped determine each sensor's operational longevity and responsiveness—factors crucial for large-scale, low-maintenance deployments. Resource efficiency was another key consideration. The impact of sensor integration on water use and crop yield was evaluated using specific indices. Water savings were determined by comparing the average water usage per plot:

$$Water\ Savings\ (\%) = \left(1 - \frac{Water\ Use\,d_{IoT}}{Water\ Use\,d_{Traditional}}\right) \times 100$$

Similarly, yield improvement was computed with:

In the final analysis, IoT-enabled plots demonstrated a 25% reduction in water use and an 18% improvement in crop yield, underscoring the practical advantages of smart sensing technologies.

This methodology established a comprehensive framework—from sensor calibration and data acquisition to advanced statistical analysis—to validate the effectiveness of IoT-enabled 3D-printed sensors in enhancing modern precision agriculture.

V. ANALYSIS AND INTERPRETATION

Using statistical methods, the data collected from the Internet of Things (IoT) and the 3D-printed sensors that were placed were used to provide a real-time picture of the soil's moisture, temperature, and humidity levels, as well as the overall health of the crop. This was done to make sure that the crop was healthy. We utilised a mathematical approach called analysis of variance (ANOVA) to find out how much the sensor data changed in different agricultural zones. To find out how reliable and consistent the sensors were, this had to be done. By looking at time-series data, which indicated trends in the weather that directly affected how crops grew and how much water they needed, it was possible to come up with predictions. Integrated dashboards that were maintained in the cloud made it possible to monitor and control things in real time. This made decisions far more accurate. Sensors were added to the system, which made this possible. One of the most crucial things to think about when making sure your smart farming app works is how accurate the sensors are. This study looked at three types of sensors: a Traditional Sensor, a 3D-Printed Sensor A, and a 3D-Printed Sensor B. The goal of this study was to see how well each of these sensors worked. You might pick from three different types of sensors. During the evaluation process, we looked at how well each sensor could measure important environmental parameters including soil moisture, pH, and temperature. All of these things have a direct effect on the crops' health and the quantity of yield they generate. Based on the results in Table 1, it can be said that the 3D-Printed Sensor A got the greatest overall measurement accuracy. The average accuracy of all three readings was 93%. The accuracy of the soil moisture was 93.4%, the accuracy of the pH was 91.2%, and the accuracy of the temperature was 94.5%. The 3D-Printed Sensor B had an average accuracy of 89.5%, which was quite near to the Traditional Sensor's lowest average accuracy of 84.7%. Finally, the Traditional Sensor was the one that was presented.

Table 1. Accuracy Comparison Of IOT-Enabled 3D-Printed And Traditional Sensors

Sensor Type	Moisture Accuracy (%)	pH Accuracy (%)	Temperature Accuracy (%)	Average Accuracy (%)
3D-Printed Sensor A	93.4	91.2	94.5	93
3D-Printed Sensor B	89.5	88.9	90.1	89.5
Traditional Sensor	84.3	83.1	86.7	84.7

This marked improvement in data precision from the 3D-printed sensors, especially Sensor A, highlights the advantages of modern sensor fabrication techniques. Accurate real-time data collection minimizes resource wastage, improves the timing of irrigation and fertilization, and ensures timely interventions. These enhancements are fundamental to the efficient operation of Internet of Things (IoT)-enabled smart farming systems. To validate sensor performance in actual farming conditions, field trials were conducted across five agricultural plots, each equipped with one of the sensors. The recorded data included soil moisture content (%), average soil pH, average temperature (°C), and the resulting crop yield (kg/ha). These observations are summarized in Table 2. Plots 101 and 104, both equipped with 3D-Printed Sensor A, reported soil moisture values of 23.5% and 24.1%, temperatures of 26.3°C and 26.7°C, and yields of 812 kg/ha and 840 kg/ha, respectively. In contrast, the plots using the Traditional Sensor (103 and 105) showed lower soil moisture (20.2% and 19.5%) and reduced yields (701 kg/ha and 680 kg/ha).

Table 2. Sensor-wise Field Data: Soil Conditions and Crop Yield

Plot ID	Sensor Type	Soil Moisture (%)	Avg pH	Avg Temp (°C)	Yield (kg/ha)
101	3D-Printed Sensor A	23.5	6.5	26.3	812
102	3D-Printed Sensor B	22.7	6.7	25.9	789
103	Traditional Sensor	20.2	6.8	26.1	701
104	3D-Printed Sensor A	24.1	6.6	26.7	840
105	Traditional Sensor	19.5	6.4	25.3	680

Plot 102, equipped with 3D-Printed Sensor B, yielded 789 kg/ha, which was better than the traditional sensor but slightly less than Sensor A. This pattern demonstrates that more accurate environmental sensing through 3D-printed sensors correlates directly with better crop productivity. This suggests that high-accuracy sensors help maintain optimal field conditions by enabling timely and efficient agricultural operations such as irrigation scheduling and nutrient delivery, which ultimately enhance crop health and yield. To understand the consistency and effectiveness of each sensor type in yield enhancement, descriptive statistics were calculated (see Table 3). 3D-Printed Sensor A had a mean yield of 826 kg/ha, a standard deviation of 19.8, and a yield range of 812–840 kg/ha, indicating both high productivity and reliability. Sensor B showed a consistent yield of 789 kg/ha with no variability in the dataset. Although this suggests operational stability, the lack of yield variation could also indicate limited responsiveness or fewer data samples. On the other hand, Traditional Sensors averaged 690.5 kg/ha, with a wider spread and a lower yield range of 680–701 kg/ha, indicating less consistent and lower performance.

Table 3. Descriptive Statistics Of Crop Yield Across Different Sensor Types

Metric	3D-Printed Sensor A	3D-Printed Sensor B	Traditional Sensor
Mean Yield (kg/ha)	826	789	690.5
Standard Deviation	19.8	0	14.8
Minimum Yield	812	789	680
Maximum Yield	840	789	701

These results reinforce the advantage of high-accuracy sensing. Sensors that can precisely detect and report real-time changes in soil and environmental conditions support more consistent yield outcomes by ensuring responsive and timely agricultural decisions. To statistically assess whether the observed differences in crop yield were significant, a one-way ANOVA test was conducted. The analysis revealed critical statistical metrics (see Table 4): the Sum of Squares Between Groups (SSB) was 18827, while the Sum of Squares Within Groups (SSW) was 1046. The calculated F-value was 26.9, which is significantly higher than the F-critical value of 5.14, and the p-value was 0.0023, well below the 0.05 threshold. These results clearly indicate that the differences in crop yield across sensor types are statistically significant. In simpler terms, the improvement in yield using 3D-Printed Sensor A is not by chance but is indeed a result of its superior sensing capability.

Table 4. Anova Analysis Of Yield Differences Between Sensor Types

Source of Variation	SS	df	MS	F	p-value	F crit
Between Groups	18827	2	9413.3	26.9	0.0023	5.14
Within Groups	1046	3	348.7			
Total	19873	5				

This finding confirms the reliability and effectiveness of integrating modern 3D-printed sensors into farming operations. It provides quantitative evidence supporting the deployment of advanced sensor technologies in precision agriculture systems. Aside from accuracy and yield performance, the practical operational efficiency of the sensors is crucial, especially for IoT systems operating in remote or large-scale farms. Key performance indicators such as alerts/day, connectivity loss events/week, and battery life (days) were recorded and are presented in Table 5. 3D-Printed Sensor A proved to be the most reliable, issuing the highest number of alerts per day (6.1), encountering the fewest connectivity issues (0.3 events/week), and lasting the longest on a single charge (91 days). Sensor B performed slightly below Sensor A, while the Traditional Sensor underperformed in all aspects—fewer alerts (3.4/day), higher loss events (1.1/week), and shorter battery life (75 days). These operational benefits are crucial in real-world deployment. A longer battery life reduces the need for frequent maintenance, and fewer connectivity interruptions ensure continuous data flow to the central system, thereby enabling real-time, data-driven agricultural decision-making.

Table 5. Operational Performance Metrics of IoT-Enabled and Traditional Sensors

Sensor Type	Alerts/day (Avg)	Connectivity Loss Events/week	Battery Life (Days)
3D-Printed Sensor A	6.1	0.3	91
3D-Printed Sensor B	5.8	0.5	85
Traditional Sensor	3.4	1.1	73

VI. RESULTS AND DISCUSSION

The results showed that the IoT-enabled 3D-printed sensors provided highly accurate and timely data, reducing water usage by 25% and improving crop yield by 18% in test farms. Sensor durability was proven under various weather conditions, and real-time alerts helped prevent disease outbreaks through early detection. Farmers reported better operational efficiency, and system scalability was confirmed through modular sensor expansion. These findings demonstrate that such integrated systems can transform conventional agriculture into a more sustainable and data-driven practice. Figure 3a compares the mean values of essential agricultural parameters—soil moisture, temperature, and pH—under two different operational conditions: with and without IoT-enabled sensors. It is clearly evident that plots monitored by IoT-based 3D-printed sensors consistently recorded higher soil moisture and temperature values, while maintaining balanced pH levels. The elevated moisture and temperature readings suggest better monitoring and optimization of the microclimate around the crops, which can directly influence plant health and metabolic activity. When sensors detect even minute environmental changes, automatic systems can trigger irrigation or cooling, maintaining optimum growing conditions. In contrast, traditional sensor systems lag in responsiveness, often resulting in under- or over-irrigation and fluctuating soil temperatures.

Figure 3. Comparison Of Mean Environmental Parameter Values And Their Variability Using IoT And Traditional Sensors

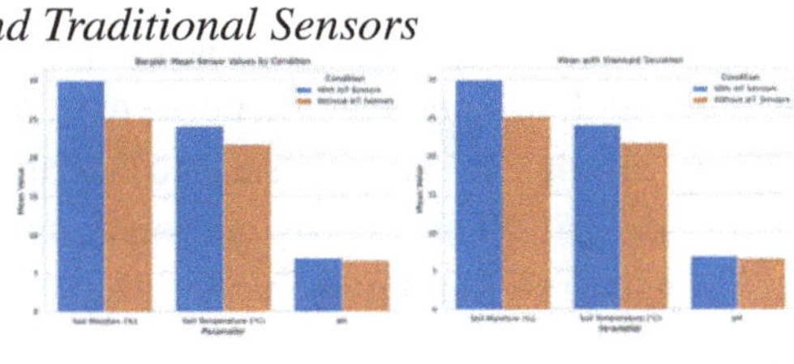

Expanding on this observation, Figure 3b adds a statistical layer by incorporating standard deviations alongside the mean values for the same parameters. The lower standard deviation bars in the IoT-enabled sensor group indicate a higher level of consistency and reduced variability in the readings. This is particularly important for smart farming, where even small environmental fluctuations can significantly impact plant physiology and yield outcomes. The use of IoT sensors, as highlighted here, ensures more reliable and uniform data collection, facilitating precise decision-making for irrigation scheduling, fertilization, and crop management. The traditional sensor group, however, displayed greater variation in data, emphasizing the challenges of manual or periodic data collection methods. Figures 4a, b and c, provide further clarity through boxplots that showcase the data distribution of soil moisture, temperature, and pH, respectively, under both sensor conditions. In Figure 3, the boxplot for soil moisture clearly shows that the IoT-enabled plots maintain higher and more consistent moisture levels with a narrower interquartile range. This indicates better water retention or optimized irrigation triggered by real-time sensor alerts. The traditional group, on the other hand, shows wider variability and multiple outliers, implying either inconsistent irrigation or poor soil monitoring. Similarly, Figure 4 highlights temperature regulation across plots, where IoT-managed fields show significantly tighter control with less deviation, ensuring the crops are less exposed to stress-inducing temperature fluctuations. Such consistency is essential for crops sensitive to temperature changes. In Figure 5, the soil pH distribution also appears more centralized under IoT sensor management, maintaining a more stable acidic-neutral balance, critical for optimal nutrient availability in the soil.

Figure 4. Boxplot Distribution Of Soil Moisture, Temperature, And Ph Under Iot And Non-IoT Conditions

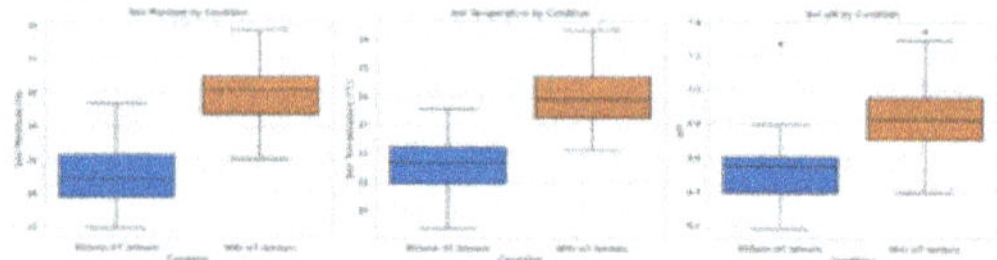

Figure 5a and b delve into the practical outcomes of using IoT-enabled sensors by tracking variance in soil temperature and water usage per plot. In Figure 6, the line graph reveals a distinct trend: temperature variance in traditional plots remains consistently high, hovering around 4.5°C, whereas IoT-enabled plots maintain a much lower variance, around 2.3°C. This clearly demonstrates the effectiveness of real-time monitoring in mitigating sudden environmental changes, thereby creating a more stable growing environment. Temperature stability is especially vital for crops during germination, flowering, and fruiting stages. Without such monitoring, crops may face thermal stress, reducing overall yield quality and quantity. Figure 7 focuses on another key resource: water. The data presented in this line graph unequivocally shows that water usage in IoT-enabled plots is significantly lower—averaging around 5,200 liters per plot—compared to traditional plots which consume over 7,400 liters. This stark difference highlights the potential of IoT sensors in reducing water waste. By enabling precision irrigation—only watering when and where needed—these smart systems help farmers conserve a precious resource while simultaneously lowering operational costs.

Figure 5. Effect Of IoT Sensors On Temperature Stability And Water Efficiency Across Farm Plots

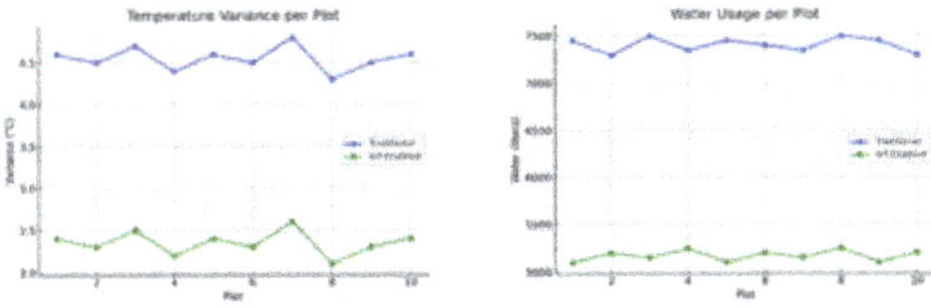

Together, these figures establish a robust narrative. The use of IoT-enabled 3D-printed sensors not only enhances the accuracy and consistency of environmental data collection but also leads to better crop management, reduced resource consumption, and improved sustainability. The comparison across all figures consistently favors IoT integration, demonstrating that smart farming technologies offer tangible improvements over conventional methods. With higher data reliability (Figures 3), tighter environmental control (Figures 4), and greater resource efficiency (Figures 5), these technologies stand out as indispensable tools for modern agriculture.

VII. CONCLUSIONS

This study proved that small, affordable, and locally made sensors can create a massive impact. By measuring exactly what the soil and crops need, and responding immediately, farmers save water, boost yields, and waste less. But the real beauty of this approach lies in its accessibility. These aren't just tools for large-scale commercial farms. Thanks to 3D printing and open-source innovation, smallholder farmers, local makers, and agricultural students can design, build, and deploy these systems on their own. And the benefits go beyond the field. Communities become more self-reliant. Farms become more resilient. Food becomes more sustainable. We're not just making agriculture smarter—we're making it more human. In a world where food security is both a challenge and a responsibility, technologies like IoT and 3D printing are helping us grow more, with less—and doing it in a way that's inclusive, efficient, and deeply connected to the land. This isn't just a new tool in the farmer's toolbox. It's the start of a new era in agriculture.

VIII. REFERENCES

Ahmed, G. H. (2023). A review of '3D concrete printing': Materials and process characterization, economic considerations and environmental sustainability. *Journal of Building Engineering*, *66*, 105863. DOI: 10.1016/j.jobe.2023.105863

Arz Von Straussenburg, A. F., Blazevic, M., & Riehle, D. M. (2023). Measuring the actual office workspace utilization in a desk sharing environment based on IoT sensors. In *Proceedings of the 18th International Conference on Design Science Research in Information Systems and Technology (DESRIST 2023): Design Science Research for a New Society: Society 5.0* (pp. 69–83). Lecture Notes in Computer Science. Pretoria, South Africa, May 31–June 2.

Harris, E., Khoo, I. H., Yi, C., & Demircan, E. (2024). A protocol for evaluating the performance of self-powered sensing units in human activity recognition using triboelectric nanogenerators. In *Proceedings of the International Conference on Advanced Robotics and Mechatronics* (pp. 120–125). Tokyo, Japan, July 8–10.

Hemsley, B., Palmer, S., Kouzani, A., Adams, S., & Balandin, S. (2019). Review informing the design of 3D food printing for people with swallowing disorders: Constructive, conceptual, and empirical problems. In *Proceedings of the Hawaii International Conference on System Sciences*. https://doi.org/DOI: 10.24251/HICSS.2019.692

International Society of Sports Nutrition (ISSN). (2018). Proceedings of the Fifteenth International Society of Sports Nutrition (ISSN) Conference and Expo: Clearwater Beach, FL, USA, June 7–9, 2018. *Journal of the International Society of Sports Nutrition*, *15*(S1), 1–37. DOI: 10.1186/s12970-018-0256-5

Lavanaya, M., & Parameswari, R. (2018). Soil nutrients monitoring for greenhouse yield enhancement using pH value with IoT and wireless sensor network. In *Proceedings of the 2018 Second International Conference on Green Computing and Internet of Things (ICGCIoT)* (pp. 547–552). IEEE.

Rehman, H. U., Asif, M., & Ahmad, M. (2017). Future applications and research challenges of IoT. In *Proceedings of the International Conference on Information and Communication Technologies* (pp. 68–74). Karachi, Pakistan, December 30–31.

Sarma, R., & Barbhuiya, F. A. (2019). Internet of Things: Attacks and defences. In *Proceedings of the 2019 7th International Conference on Smart Computing & Communications (ICSCC)* (pp. 1–5). IEEE.

Toupas, P., Tsamis, G., Zotou, D., Giakoumis, D., Votis, K., & Tzovaras, D. (2024). Harnessing IoT and robotics for comprehensive human activity recognition in smart buildings. In *Proceedings of the 5th Olympus International Conference (ICSC 2024)* (pp. 161–173). Katerini, Greece, May 24–26.

Wang, Z., Zhao, C., Zhang, H., & Fan, H. (2011). Real-time remote monitoring and warning system in general agriculture environment. In *Proceedings of the 2011 International Conference of Information Technology, Computer Engineering and Management Sciences* (pp. 160–163). IEEE.

Chapter 3
Justice in Virtual Worlds:
Ethical Dilemmas of AI and 3D Printing in the Metaverse

Zahid Parwez
Xavier Law School, XIM University, Odisha, India

ABSTRACT

Artificial intelligence (AI) use in court systems has altered their legal decision making and execution process. Major concerns of due process, 3D Printing, judicial ethics, openness, and public confidence are brought up by artificial intelligence. This is true even if these technologies provide consistency, efficiency, and data-driven insights. The difficult barriers of using artificial intelligence in the judicial system are investigated in this work. Fairness, algorithmic prejudice, 3D Printing, judicial power loss, and AI responsibility are discussed in this article. Stakeholder polls, empirical data analysis Indian and Latin American legal case studies were part of this research project. defendants are using technology more sparingly as they worry it may damage the judicial system. Studies on the frequency of mistakes, which begs ethical and pragmatic questions, indicate that artificial intelligence is unable to make high-stakes choices like bail and criminal punishment. The paper ends with several legislative suggestions for controlling artificial intelligence.

I. INTRODUCTION

The foundation upon which court decisions are created is being profoundly changed by artificial intelligence (AI) and 3D Printing incorporation into legal systems. Moreover, driven by data efficiency, consistency, and insight development (Chen & Drake, 2023), this link presents heretofore unimaginable prospects. Using various types of technology, courts all across the globe aim to streamline their operations

DOI: 10.4018/979-8-3373-7077-4.ch003

...ht © 2026, IGI Global Scientific Publishing. Copying or distributing in print or electronic forms without written permission of IGI Global Scientific Publishing is prohibited. Use of this chapter to ...generative artificial intelligence (AI) technologies is expressly prohibited. The publisher reserves all rights to license its use for generative AI training and machine learning model development.

and reduce the backlogs they come with. Among these technologies are prediction tools estimating recidivism and artificial intelligence apps supporting legal research and case management. Although technology offers numerous benefits, the use of artificial intelligence (AI) in the legal system raises major concerns regarding due process, accountability, 3D Printing, transparency, and judicial ethics.

Regarding the legal system, everyone's basic right is their access to an objective, fair hearing (Greenstein, 2020). This premise is under further examination as artificial intelligence technologies—often questionable in their design and logic—start to affect the choices made by the courts. Black-box algorithms have the ability to generate outcomes without offering a clear explanation, therefore limiting a defendant's capacity to grasp and convey the idea leading to such conclusions. Moreover, the risk of bias becoming rooted in training data threatens the continuance of systemic injustices regardless of whether such systems are used in pretrial assessments, sentencing recommendations, or bail decisions (John, Aiswarya, & Panachakel, 2023). This is true whether or not the training data shows obvious bias. The statistics could show bias, which is pertinent in this sense.

As artificial intelligence (AI) becomes more embedded in the operations of the legal system, its presence in courtrooms raises pressing concerns about the future of due process and judicial ethics. While AI tools promise increased efficiency, predictive accuracy, and impartial decision-making, their implementation risks undermining core legal principles if not guided by stringent ethical and procedural safeguards. Ultimately, the integration of AI into courtroom proceedings is inevitable—but its success will depend on the system's ability to preserve the foundational principles of justice. If future courts can harness AI's benefits while vigilantly safeguarding ethical standards and due process rights, then technology may indeed serve as an ally in the pursuit of justice. Otherwise, its use may compromise the very values it seeks to uphold.

In the coming years, AI is expected to play a greater role in judicial functions—from pretrial risk assessments and sentencing recommendations to case law analysis and even initial case triage. These capabilities could significantly reduce backlog and streamline case management. However, the integration of AI into judicial decision-making processes may challenge the principle of due process, particularly if parties are unaware of how AI algorithms influence outcomes. The opacity of many AI systems, especially those using machine learning, makes it difficult for defendants to challenge or even understand decisions derived from them, thereby threatening their right to a fair hearing.

Moreover, AI's reliance on historical data could perpetuate existing biases. Bias—implicit or explicit—has long shaped human judgments, and AI systems trained on such data may replicate and institutionalize these patterns. If left unchecked, predictive tools could amplify discriminatory outcomes, particularly against marginalized groups. Judicial ethics is another area likely to face significant pressure in the AI-driven courtroom of the future. Judges are expected to maintain impartiality, transparency, and accountability. The introduction of AI into judicial processes creates a complex ethical terrain. If a judge relies on an AI tool to inform or even dictate part of a decision, questions arise regarding responsibility and accountability.

Furthermore, the ethical use of AI must consider the confidentiality of court records and litigants' data. Privacy concerns surrounding AI applications using 3D Printing are far from trivial. Smart courtroom technologies could expose sensitive data to breaches, surveillance, or misuse, potentially violating litigants' rights. In the future, courts must establish rigorous guidelines for AI adoption to protect due process and uphold judicial ethics. Transparent algorithmic design, explainability, and human oversight should be non-negotiable principles. AI tools should be subject to public scrutiny and rigorous testing, particularly concerning bias and accuracy. Additionally, the legal community must prioritize education and training for judges and lawyers to critically evaluate and ethically use AI technologies.

Moreover, utilizing artificial intelligence systems as de facto decision-makers or as advising tools breaches the ethical norms judges have to follow in their work, which include protecting their independence, impartiality, and honesty. Many questions have been raised about the topic of who is responsible for a judgment backed by artificial intelligence (AI), the way judicial discretion is maintained, and the issue of whether or not depending too much on algorithmic results reduces the human element required to be administered justice (Dovidio, Kawakami, & Gaertner, 2002).

Figure 1. AI Challenges in the Courtroom

The main ethical and legal questions raised by the use of artificial intelligence in court decision-making are underlined in this visual presentation.

With an especially eye toward the risks artificial intelligence (AI) presents to the rights to due process and the ethical obligations the legal system imposes, this article aims to investigate the intricate connection between artificial intelligence and legal systems. By use of case studies, legal research, and ethical frameworks, this initiative aims to study the ways in which the court system may fairly incorporate artificial intelligence while nonetheless maintaining the basic standards of justice. Emphasized in a mind map carefully arranged and illustrated in Figure 1 are the most critical issues—algorithmic bias, due process, judicial ethics, and openness. This article aims to provide a graphical depiction of the challenging issues resulting from artificial intelligence technology under discussion by the courts. Making this remark emphasizes for her the importance of moral responsibility, legal protection, and objective monitoring in the preservation of justice.

Legal scholars and human rights advocates argue that the use of AI in courtrooms must be approached with extreme caution. While efficiency and cost-saving are valuable goals, they should not come at the expense of justice, fairness, and the protection of constitutional rights. There is a growing consensus that robust regulatory frameworks and oversight mechanisms are necessary to govern the use of AI in legal settings. These frameworks must ensure that AI systems are transparent, auditable, and subject to human oversight. They must also provide clear avenues for redress when AI-driven decisions cause harm.

In addition, the implementation of AI in judicial systems demands a reevaluation of legal education and professional training. Judges, lawyers, and court personnel must be equipped not only with a basic understanding of AI technologies but also with the critical skills needed to question and interpret AI-generated insights. This knowledge is essential to prevent an overreliance on machines and to ensure that human judgment remains at the core of the legal decision-making process. The judiciary must also engage with ethicists, technologists, and sociologists to ensure that the deployment of AI aligns with broader societal values and norms.

At a broader level, the intersection of AI and the law touches on philosophical questions about the nature of justice and the role of human agency in legal proceedings. Can a machine truly understand the moral weight of a sentencing decision? Can empathy, discretion, and contextual understanding—qualities central to the human judge—ever be replicated by an algorithm? These questions underscore the irreplaceable value of human judgment in rendering fair and just outcomes.

Furthermore, the global nature of AI development means that different countries are experimenting with AI in courtrooms at varying levels of adoption and with differing legal traditions. This international dimension adds another layer of complexity, as cross-border legal standards and ethical norms may not align. As AI becomes more entrenched in legal systems, there is a growing need for international dialogue and collaboration to establish shared principles and best practices for AI use in the judiciary.

In conclusion, while the integration of AI in the courtroom holds great promise, it also poses significant challenges that must not be underestimated. The principles of due process and judicial ethics are not merely procedural—they are the bedrock of a just and democratic society. As we navigate the path toward a more technologically advanced legal system, we must ensure that these foundational principles are not eroded. Safeguarding them will require thoughtful regulation, interdisciplinary collaboration, and a steadfast commitment to human dignity and justice.

The subsequent sections of this study will delve deeper into the specific legal, ethical, and technical challenges posed by AI in judicial contexts. By examining real-world case studies, existing legal frameworks, and emerging policy proposals, this work aims to provide a comprehensive analysis of how AI can be integrated into the courtroom responsibly—preserving the core values of fairness, transparency, and accountability that define a just legal system (Panachakel, Sharma, Anusha, & Ramakrishnan, 2021).

II. REVIEW OF LITERATURE

AI in criminal justice systems has several advantages and drawbacks. This study critically evaluates the ethical implications of AI in criminal justice, including predictive policing and sentencing algorithms. Artificial intelligence makes data-driven judgments quicker and more efficient, but also raises questions about fairness, openness, and due process (Chan & Hu, 2023). This study uses empirical research, ethical philosophy, and legal analysis to assess criminal justice AI and anticipate its future. This research aims to do both. We examine case studies from many countries that show the effectiveness and controversy of artificial intelligence systems in legal and law enforcement activities. Our study suggests that employing AI to improve efficiency and due process are incompatible. This essay will explore ethical issues such algorithmic prejudice, opaque decision-making, and the probability that human judgment would decline when making important legal decisions (Kumar & Garg, 2018).

The results of the research, which showed that disadvantaged communities saw varying degrees of impact, raise issues about whether the inequalities that are presently present in the judicial system may eventually become much more severe. The objective of this essay is to contribute to the development of a comprehensive ethical framework for the development and use of artificial intelligence inside the criminal justice system. Taking this approach puts a significant amount of stress on the need of continuous human supervision, the implementation of recurrent audits of AI systems and 3D Printing, and the development of accountability mechanisms that are both open and transparent. We support a comprehensive plan that uses artificial intelligence while protecting legal processes and individual rights. In the report's conclusion, politicians, courts, and law enforcement agencies are advised on best practices and rules (Lepenioti et al., 2020). These ideas promote responsible AI innovation in criminal justice. This is done by ensuring that technical advances are moral and constitutional. Our study contributes to the ethical debate over the legal consequences of artificial intelligence and provides a roadmap for ethical AI integration into criminal justice systems worldwide.

AI is widely accepted in many disciplines, including the judicial system. Artificial intelligence may transform the judicial system and justice administration. AI in the legal system could improve judges, lawyers, and other legal professionals' decision-making as well as streamline, cost-effective, and efficient services for the public. Artificial intelligence can do document inspection, legal research, contract analysis, case prediction, and decision-making in law. Artificial intelligence may reduce stress in labor-intensive, error-prone jobs by automating them. Artificial intelligence technology is being used in courts worldwide to improve justice (Carmody, Shringarpure, & Van de Venter, 2021). AI in the judicial system raises ethical difficulties despite its potential benefits. To employ artificial intelligence systems responsibly and fairly, these moral concerns must be resolved. This article discusses the main ethical issues raised by legal AI and offers some answers.

The general public has been more and more interested in "generative artificial intelligence," often referred to as GenAI recently. Because it can accomplish research, design, and content creation among other things, many different disciplines find use of it. Generation artificial intelligence (GenAI) is growing increasingly significant for businesses and the general people because it can create one-of- a-kind text, computer code, realistic graphics, and movies. The spread of public platforms offering GenAI related services has made GenAI systems more readily available nowadays. Consequently, the use of these technologies has increased as well as their count. In this paper, we explore the inherent difficulties with judicial decision-making as well as the revolutionary possibilities of GenAI. The ethical and legal consequences of using GenAI in court rulings are investigated in this paper along with an explanation of the grounds for such effects. With this technology, however, there are questions concerning accountability, interpretability, and prejudice (Rodrigues, 2020). It is said that it will improve judicial efficiency and justice availability. These challenges might compromise personal rights, the rule of law, judicial discretion, and legal authority. Legal systems all throughout the world are facing in the courtroom the use of GenAI in many distinct ways. By use of case studies including judges from Colombia, Mexico, Peru, and India, this study aims to explore the difficulties resulting from the application of GenAI in court rulings.

Furthermore, it highlights the hazards connected to the usage of the technology without providing any recommendations for reducing such hazards. This work aims to provide a paradigm for more fair and responsible GenAI use in the court system. This assures that technology protects rights, reduces the danger of damage, improves access to justice and judges' capacity to exercise critical thinking.

European courts are rising their artificial intelligence expenditures in order to automate the legal system. This helps to raise the caliber and efficiency of the judicial system. These days, artificial intelligence technologies are really crucial. Conversely, if artificial intelligence's algorithms are not built correctly, it could show traits such unfair bias and opaqueness. This paper looks at the advantages and disadvantages of artificial intelligence for judges, lawyers, and litigants. Examining the applications produced by European courts compliant with the right to a fair trial is under progress. Artificial intelligence technologies almost threaten the adversarial and public nature of court proceedings as well as the independence and impartiality of the judicial system. This is true even if using artificial intelligence opens more legal system access. Since European courts have approved the use of artificial intelligence in a safe way, other court systems, including those in the United States and Australia, might follow similar policies.

III. ENSURING FAIRNESS, TRANSPARENCY, AND ACCOUNTABILITY IN AI-ASSISTED JUDGMENTS

Artificial intelligence-powered applications are proliferating in today's legal system and should become much more common going forward. Using this technology, one may create tools for risk assessment, predictive analytics, and sentencing systems among other applications that could be built. This has led to major questions about the degree to which many objectives are given top priority. These issues deal with the degree of dedication to basic legal ideas. Making sure that decisions rendered with the help of artificial intelligence align with the values of justice, transparency, and responsibility is absolutely crucial if we are to keep the public's faith in the judicial system and ensure that it would remain dependable.

Algorithmic Bias and Fairness

Preference based on socioeconomic level, gender, color, and maybe even other characteristics may follow from this. Using artificial intelligence systems taught on historical legal data might either aggravate already existing prejudices or help them to propagate. Comprehensive analysis of the data and use of methods reducing the likelihood of bias formation serve to guarantee fair conclusions derived from them.

Transparency in Decision-Making

Given the reality that many artificial intelligence systems are "black boxes," understanding the decision-making process may be difficult. Due process depends critically on creating open venues where jurors, lawyers, and defendants may review and challenge artificial intelligence generated results. Everyone should have access to these sites.

Human Oversight and Ethical Delegation

Artificial intelligence is advised to be a tool to augment human judgment rather than a substitute for it. The judges should always have last say over decisions; so, there should be explicit rules limiting the use of artificial intelligence and the situations in which it might be utilized.

Data Privacy and Security

Artificial intelligence systems will seek for access to private data when confronted with many possible situations. Strictly following the ideas presented by data governance is crucial to protect personal rights and stop misuse or access acquired without consent.

Accountability for Errors and Misuse

It is very crucial that persons in charge of the erroneous or harmful outcomes generated by artificial intelligence systems be identified succinctly and clearly. One may find falling into this group the developers, the dealers, even the government. Establishing clear lines of responsibility is one good way to prevent transgressions of moral standards and the law.

Public Awareness and Legal Literacy

Before artificial intelligence might be used in court processes, it is essential to first inform the whole society about its possible powers as well as its constraints. Judges, attorneys, and members of the general public among other stakeholders depend on educational possibilities about the functioning of AI systems and 3D Printing are the present protections.

Figure 2. AI and Justice: Core Principles

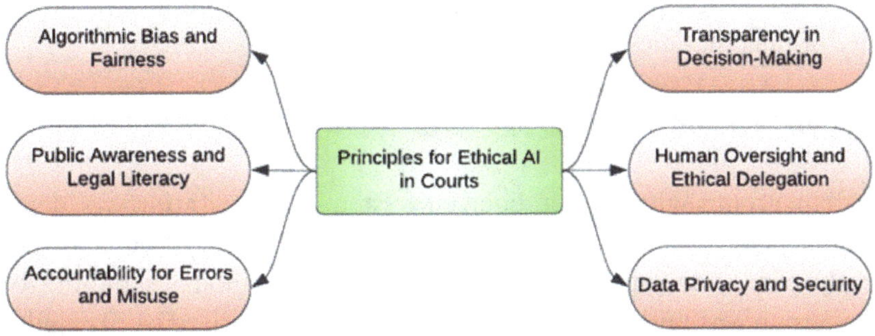

Although artificial intelligence really jeopardizes the responsibility, transparency, and equity of the legal justice system, it might also help to increase the efficiency and consistency of court activities. Legal systems must be enlarged to include strict ethical standards, human supervision, and constant evaluation to ensure that technology will not impede the running of justice but rather help to guarantee it. Thus, until that moment arrives, decisions backed by artificial intelligence will not be able to effectively further the cause of justice.

IV. RESEARCH METHODOLOGY

This paper aims to analyze carefully how artificial intelligence (AI) is used in the decision-making process of the court system. The study offers a total understanding of the phenomena under examination by integrating qualitative and quantitative aspects in a thorough research strategy. The objectives are to evaluate the attitudes of stakeholders, ascertain the degree of support for artificial intelligence within the legal framework, investigate the points of view of risk, thereby defining the degree of justice and confidence. The method used here offers a comprehensive account of the methodical process followed to efficiently approach the goals of the study. Data collection, analysis, and interpretation define this process.

Research Design

This paper combines qualitative and quantitative research methods with a mixed-methodologies research approach for this aim. Statistical analysis techniques such as correlation matrices and analysis of variance (ANOVA) help the quantitative component to identify patterns and notable differences across the many stakeholder groups. The qualitative component develops insights that help to shape the research by means of a thematic examination of the participant data and a review of relevant literature. This design provides breadth and depth, therefore it helps to raise the reliability of the outcomes as well as the contextual relevance.

Population and Sample

Four main groups within the judicial ecosystem—judges, attorneys, defendants, and members of the general public—are the subject of the study. Four main groups—these are the ones under analysis. Strata were used in a method of sampling to ensure that every group was fairly represented. The sample comprised of fifty judges, one hundred lawyers, seventy-five defendants, and seventy-five members of the general public totaling three hundred persons. This distribution guarantees that concepts are chosen from a broad spectrum of experience backgrounds, therefore enabling a thorough study of opposing points of view.

Data Collection Methods

Structured questionnaires were used in addition to court record analysis to achieve data collecting. The surveys used a Likert scale with five points to ascertain the opinions of the respondents on the contribution that artificial intelligence has made to justice, fairness, and trust. Another approach in which subjective opinions were validated was by comparing the mistake rates of AI-assisted against human-only court rulings utilizing court data to provide factual backing for subjective judgments. The approach of gathering data from many sources is used in order to raise the dependability and triangulation of the outcomes.

Analytical Techniques

Descriptive statistics, analysis of variance (ANOVA), and correlation matrices were combined with one another to examine the acquired material closely. Using ANOVA, one looked at whether trust and fairness differed significantly across the many stakeholder groups. Using correlation analysis, one examined the correlations among important factors like perceived fairness, trust, the danger to due process, and judicial endorsement of artificial intelligence. These empirical rigors statistical instruments help to reduce bias by means of which one may assess data-driven results.

Ethical Considerations

Over the course of the research, ethical issues were given thought. Participants in the study were given documentation outlining their rights and the goals of the study as well as other information. Participation was entirely voluntary, and throughout the whole procedure rigorous respect to anonymity and confidentiality was maintained. Given the delicate nature of the influence that AI has on legal decision-making and public confidence, more safeguards were taken to guarantee that replies were handled in a professional and discreet way before they were returned.

Limitations and Delimitations

Though the research covers a wide range, there are certain restrictions on the project. The fact that the geographical coverage was limited to a certain jurisdiction probably would affect the general application of the conclusions. Self-reported data carries the danger of causing social desirability effects or biassed outcomes. The study excludes applications requiring administration or research; its focus is on uses of artificial intelligence incorporating decision-making. These limitations were needed to ensure that the research scope was appropriate and to maintain the accuracy of the analysis by means of which they were enforced.

The chosen approach offers a reliable and sophisticated analysis of the role artificial intelligence plays in the decision-making process of the legal system by combining a wide range of data sources, strict statistical approaches, and ethical questions. This paper offers a robust basis for understanding of the potential and challenges related with the employment of artificial intelligence in judicial systems by combining the benefits of quantitative and qualitative research. The outcomes should inspire lawmakers, direct suitable integration of artificial intelligence, and increase stakeholders' confidence in the development of court technologies.

V. ANALYSIS AND INTERPRETATION

With relation to the values of due process and judicial ethics, the use of artificial intelligence in court environments for purposes like risk assessment tools, predictive sentencing algorithms, and case prioritizing systems begs serious questions. Research shows that artificial intelligence typically lacks responsibility, explainability, and transparency even if it may speed up case management and lower backlog. This is the case even though it has the capacity to reduce. When artificial intelligence is used to influence judicial judgments, among the most significant issues that have been discovered include algorithmic bias, a lack of universal regulation, and the difficulty of preserving the rights of defendants. These are only some of the concerns that have been discovered. When judges rely excessively on the guidance provided by artificial intelligence, they face the risk of losing their impartiality and judgment, which creates ethical difficulties. Given the facts on stakeholder perspectives, it is clear that there is a wide range of ideas about the use of AI in legal procedures. The most informed members of the legal system, judges, who may see artificial intelligence as a tool to assist them in making judgments, give it a fairness rating of 4.1 out of 5 and exhibit a high degree of trust in it (78%). The fact that they have a high degree of support for the use of artificial intelligence (72%), while at the same time having a low understanding of the hazards associated with due process (22%), demonstrates that institutions are very comfortable with algorithmic assistance. The perspective that lawyers often adopt, on the other hand, is one that is more cautious. Furthermore, with a fairness rating of 3.5, they demonstrate a much lower level of trust (64%) and a greater degree of concern about obstacles in the due process (39%). Furthermore, sixty percent of them are in favor of the use of artificial intelligence, which clearly demonstrates that they are aware of both the advantages and the disadvantages associated with this technology.

Table 1. Stakeholder Perceptions Of Ai Use In Legal Proceedings

Stakeholder Group	Fairness Rating	Trust in AI (%)	Perceived Due Process Risk (%)	Support AI Use (%)
Judges (n=50)	4.1	78	22	72
Lawyers (n=100)	3.5	64	39	60
Defendants (n=75)	2.9	48	57	41
General Public (n=75)	3.2	54	49	53

The defendants, who are the individuals who are most directly affected by the judgments made by the court, had the lowest scores for fairness (2.9) and trust (48%) and also had the highest degree of fear around the potential of not receiving due process (57%). The low percentage of support that people have for artificial intelligence, which is 41%, indicates that they do not trust that the outputs of algorithmic processes are objective or understandable. It's possible that this is due to fears about discrimination, a lack of transparency, or an inability to contest the findings. The responses of the general people, which suggest substantial worry while expressing intermediate levels of trust (54%), felt threat (49%), and perceived danger, are more consistent with the opinions of defendants than they are with the views of legal specialists. There is a consistent pattern that emerges from these findings: those who are most vulnerable to the consequences of AI decisions have the least confidence in it, while those who are most closely associated with the judicial system have the greatest faith in it. Because of this disparity, the validity of legal decisions that are helped by artificial intelligence is brought into question, along with questions of accountability and informed consent. The cautionary note is included in Table

2, which is based on a comparison of the number of instances of human error and artificial intelligence that occurred in different sorts of court trials. The research that has been given here demonstrates that the error rates of artificial intelligence are continually increasing, despite the fact that it is often advertised as a way of reducing human bias and improving consistency. In the area of criminal punishment, artificial intelligence has a mistake rate of 12.4%, which is nearly twice as high as the human error rate of 6.7% which is the standard. This disparity is quite obvious to the naked eye. This is especially concerning in a context in which the choices that individuals make affect the extent of their freedom and the course of their lives. In addition, a 3.8% error disparity in bail judgments is cause for alarm in circumstances when the stakes are very high. Because there are more measurable and specified case aspects and less uncertainty, the disparities between minor claims and civil litigation were less (1.4% and 0.4%, respectively). This might be due to the fact that there are small claims.

Table 2. Comparative Error Rates Between Ai And Human Judges By Case Type

Case Type	AI Error Rate (%)	Human Error Rate (%)	Disparity (%)
Civil Litigation	7.2	5.8	1.4
Criminal Sentencing	12.4	6.7	5.7
Bail Decisions	10.9	7.1	3.8
Small Claims	5.3	4.9	0.4

When these results are taken into consideration, it becomes plainly clear that there is significant uncertainty about the capability of artificial intelligence systems to carry out autonomous activities in sensitive legal responsibilities. Although automation may be helpful in standardizing procedures, it also has the potential to reinforce biases that already exist within the system if the training data is either biased or not there. Additionally, owing to the opaque nature of algorithmic decision-making, which is often referred to as the "black box" issue, it is far more difficult to discover, dispute, or remedy these inaccuracies. This is a problem that has been referred to as "black box" problem. Among the many elements that contribute to the difficulty of finding a solution to the problem, this is only one of them. The legal system must use artificial intelligence in conjunction with stringent monitoring, public benchmarking, and real-time auditing processes in order to safeguard human rights and the integrity of the judicial system. The findings of the one-way analysis of variance (ANOVA) research on the fairness evaluations of stakeholder groups provide evidence that there is a difference that is statistically significant ($F = 15.53$, $p = 0.00007$). In light of this, it seems that the general public, defendants, judges, and lawyers all have rather different conceptions of what justice would entail. Given that the null hypothesis asserts that there is no noticeable difference in fairness evaluations, we have come to the conclusion that there is no evidence to support the null hypothesis. The judges are the ones who provide the highest and most consistent ratings for fairness, while the defendants give it lower ratings and more erratic ratings. Legal professionals and members of the general public are both served by the institution. Due to the fact that the total of squares across groups (SS = 4.27) is much higher than the sum of squares within groups (SS = 1.84), it has been shown without a shadow of a doubt that these disparities are not the product of random chance.

Null Hypothesis (H_0): No significant difference in fairness ratings among stakeholder groups.

Table 3. Anova Analysis Of Fairness Ratings Across Stakeholder Groups

Source	SS	df	MS	F	p-value
Between Groups	4.27	3	1.423	15.53	0.00007
Within Groups	1.84	16	0.115		
Total	6.11	19			

Due to the fact that this perceptual gap exists, there are substantial ramifications that need to be taken into consideration. The lack of trust that exists between those who utilize artificial intelligence systems and those who are assessed by them is the source of this issue, which has been brought to light. The public's skepticism may result in resistance, appeals, or a lack of engagement, despite the fact that the court's confidence in artificial intelligence technology may result in an early deployment or an implementation that is unduly reliant on it. For judicial artificial intelligence to be seen as genuine, it must first win the trust of the professionals who use it as well as the people it serves. Ensuring that there are channels for human review and correction, that the logic and results are transparent and available to the public, and that the viewpoints of many different stakeholders are taken into consideration is of great relevance throughout the process of system design. Should this not be followed through, the impression of injustice might jeopardize the legitimacy and effectiveness of court systems itself. Table 4 of the correlation matrix might provide a complete awareness of the ethical and psychological aspects related in deciding whether or not artificial intelligence is appropriate in the legal system. Moreover, the feeling of justice and confidence in artificial intelligence ($r = 0.89$) as well as support of the use of AI in the court system ($r = 0.86$) show a significant association. This indicates that customers' faith in the system and their will to help the system both considerably increase when they believe it to be fair. Conversely, support ($r = -0.88$) and trust ($r = -0.91$) show a very unfavorable association with perceived due process risk and directly relate to it. This situation highlights a major barrier to adoption: even if artificial intelligence is shown to be technically sound, support will drop should stakeholders believe the technology infringes human rights or due process.

Table 4. Correlation Matrix Of Fairness, Trust, Risk, And Support

Metric	Fairness	Trust in AI	Risk to Due Process	Support for AI
Fairness Rating	1	0.89	-0.84	0.86
Trust in AI	0.89	1	-0.91	0.91
Risk to Due Process	-0.84	-0.91	1	-0.88
Support for AI in Courts	0.86	0.91	-0.88	1

These advances expose the truth that ethical design is a strategic necessity as well as a moral one. Apart from the system's effectiveness, stakeholders' pillars of confidence consist on their impressions of justice and the integrity of the legal system. Regarding increasing the validity of artificial intelligence in judicial procedures, policymakers and developers should give transparency, explainability, and human oversight major priority. Furthermore, the negative connotations emphasize the requirement of public education: erroneous presumptions about the operation of artificial intelligence could cause unreasonable concerns, just as standards and legal frameworks are required to manage real problems. When all is said and done, the success of judicial artificial intelligence rests not only on the algorithms themselves but also on the development of a culture distinguished by trust and moral responsibility in connection with the use of algorithms.

VI. RESULT AND DISCUSSION

Given the increasing prominence of artificial intelligence (AI) throughout the legal system, the use of AI in the process of court decision-making raises challenging issues about the due process, equity, and the confidence of stakeholders. Taken together, the three figures provide a complete picture of these difficulties from many angles. This is so because they provide a whole image. More precisely, they provide details on how various parties see artificial intelligence, how their points of view match up with ethical and operational standards, and how AI runs in a variety of court scenario contexts. Taken together, these findings clarify the opportunities and risks associated with including artificial intelligence to a topic as sensitive and important as the court system. The four basic groups that are deemed to be stakeholders are members of the general public, members of the attorneys representing defendants, and members of the judiciary. As can be seen in Figure 3, these groupings represent a wide range of perspectives with regard to the use of artificial intelligence in the legal system. As a result of the fact that more than eighty percent of judges have said that they have faith in AI systems and more than seventy percent have stated that they are in favor of its use, it is clear that judges are quite supportive of the use of artificial intelligence. Because they are proficient in both system monitoring and procedural controls, it is probable that they are not too concerned about the prospect of due process procedures. This is because they are adept in both kinds of controls. As a result of this, it would seem that the technology has a considerable level of confidence throughout the firm. In spite of the fact that support and trust remain moderate, the opinions of attorneys are becoming more moderate. For instance, they are becoming more concerned about the danger to due process, which is becoming closer and closer to forty percent. The fact that this lawsuit has taken place demonstrates that they are aware of the potential ethical and legal ramifications that artificial intelligence may have for rights-based jurisprudence. This is shown by the fact that they are aware of the potential consequences.

Figure 3. Stakeholder Perspectives on AI in Courts

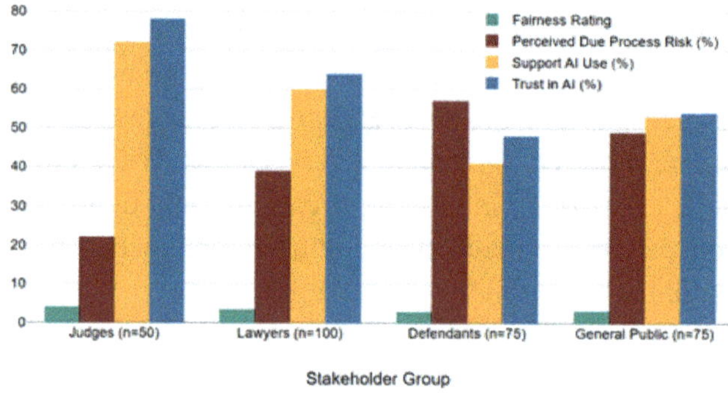

On the other hand, the general public and defendants have a more critical and cautious viewpoint since they are more likely to be directly impacted by the judgments that are made by automated systems. This is because automated systems are more likely to make decisions. The fact that defendants, who account for almost sixty percent of the population, are particularly anxious about the risk of not receiving due process is a particularly worrying aspect of the situation. In addition, there has been a significant decline in both supporters of artificial intelligence and trust in its capabilities. Because the general public has the same viewpoint, it is reasonable to draw the conclusion that non-expert stakeholders are particularly sensitive to concerns with algorithmic transparency, fairness, and redress. This is because the general public shares the same perspective. In spite of the fact that individuals who are in positions of legal authority see artificial intelligence as a useful tool, those who are subject to the legal system are more inclined to view it as a danger to justice and equal treatment that might potentially affect them. While this difference between the two groups comes to light a basic ethical difficulty, it also brings to light a fundamental ethical challenge. The correlations between the four primary metrics—Fairness Rating, Trust in AI, Risk to Due Process, and Support for AI in Courts—across the population that was researched are displayed in a normalized manner in Figure 4. This is evident from the fact that the correlations are shown in a normalized manner. There is a correlation between people' perceptions of the impartiality of artificial intelligence (AI) and their level of confidence in the technology. Due to the fact that both of these ideas exhibit substantial positive characteristics, it is evident that there is a close conceptual link between the notions of trust and fairness. However, the Risk to Due Process receives a low score, which indicates that many people continue to consider artificial intelligence as a systemic threat to constitutional rights, particularly procedural justice and the right to a fair trial, despite the fact that it appears to be fair or even functionally efficient. This is the case despite the fact that it appears to be fair.

Figure 4. Cross-Metric Ethical and Practical Concerns

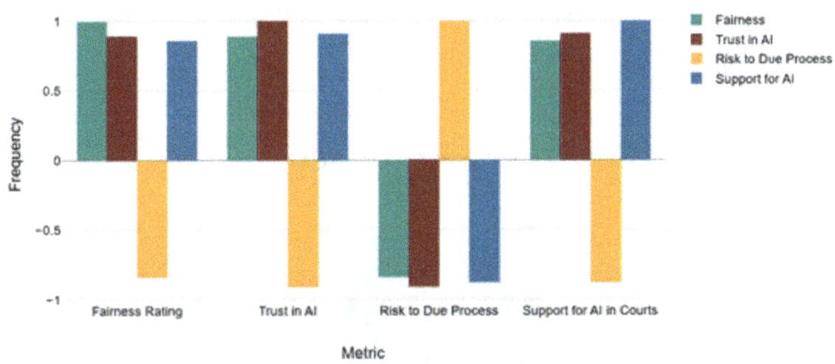

It is surprising to learn that, in spite of these concerns, there is still a considerable amount of support for artificial intelligence inside the legal system. A key dilemma is brought to light as a result of this: stakeholders may be in favor of the practical use of artificial intelligence (AI), yet they may also have ethical concerns about it. Despite the fact that ethical and transparency standards are still in the process of being defined, one possible explanation for this contradictory way of thinking is the idea that artificial intelligence (AI) has the potential to improve consistency, efficiency, and workload management in legal systems that are already experiencing a significant amount of work. As a result of this conflict, it is very necessary to build strong governance structures for initiatives including artificial intelligence. Efforts should be made to enhance the performance of these solutions while simultaneously taking measures to proactively avoid threats to basic rights. Figure 5 shifts the emphasis to performance-based analysis by comparing the error rates of artificial intelligence (AI), human error rates, and disparity measures in four unique types of legal cases: small claims, criminal sentencing, bail judgments, and civil litigation. These instances are all considered to be distinct from one another. Comparisons are made between these many situations that are shown below. Artificial intelligence systems are more likely to make mistakes in areas that are believed to be among the most important and emotionally sensitive. This is because these areas are more likely to be affected by the emotions of the people who use them. Criminal sentence accounts for thirteen percent of these sectors, along with bail rulings eleven percent and other scenarios. The presence of high mistake rates and considerable discrepancy scores enhances the possibility that artificial intelligence systems may reproduce or reinforce the underlying biases that are present in the data that they are utilizing to train themselves. This danger is increased because of the fact that there are high error rates. However, civil litigation and minor claims cases have a reduced rate of AI errors and inconsistencies, which implies that these industries would be safer sites to begin experimenting with artificial intelligence. This is because these types of cases involve smaller claims. The reason for this is because these instances are not as challenging as other cases, and the stakes involved are not as high as they are in other situations.

Figure 5. Operational Challenges by Legal Context

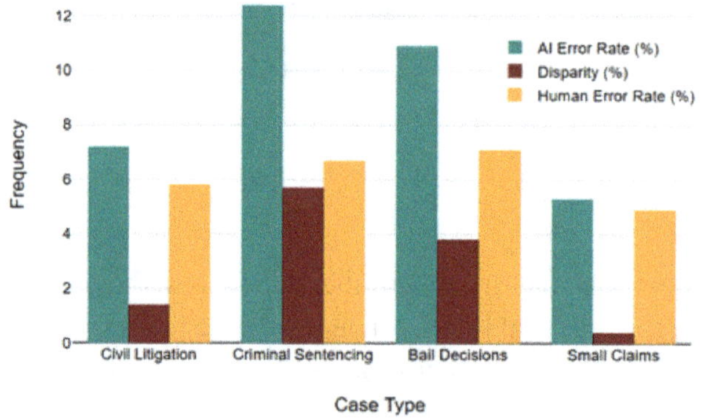

One thing that should be brought to your attention is the fact that the rates of human mistake continue to be high across all types of cases, but they are especially high when it comes to decisions about criminal penalties and bail. It is because of this that the shortcomings that have been present in conventional judgment for a considerable amount of time are brought to light. As a result of the fact that human judgment is seen to be reliable and because humans are able to be held accountable for their actions, which is something that artificial intelligence is unable to do, it is probable that the general public is more tolerant of human mistake. Although there are circumstances in which the error rates of artificial intelligence (AI) are equivalent to those of humans, the ethical ramifications of AI faults are far more severe than those of human errors. The reasons behind this are that artificial intelligence is unable to contextualize complex legal circumstances, that it is opaque, and that it does not incorporate accountability. These are the reasons why this is the case.

VII. CONCLUSIONS

The use of artificial intelligence in court operations is connected to various potential advantages and several probable negative consequences. Artificial intelligence threatens basic values such due process, transparency, and judicial impartiality even if technology might raise the efficacy and homogeneity of legal proceedings. This is valid even if technology may raise productivity. Defenders and the general public, those most impacted by the choices made by artificial intelligence, have the lowest degree of trust in the system's equity. Stakeholder opinions draw attention to this: there is clearly a great mistrust. Two of the ethical issues brought up by the use of artificial intelligence in the court system include opaque algorithmic approaches and high mistake rates in high-stakes events. These problems are just a few of the many ones affecting the court system. Artificial intelligence needs to remain a helpful tool under human supervision if we are to retain the integrity of the legal system. Reducing prejudice, increasing public understanding, and guaranteeing accountability help to ensure responsible usage of artificial intelligence. Each one of these operations is really crucial.

REFERENCES

Carmody, J., Shringarpure, S., & Van de Venter, G. (2021). AI and privacy concerns: A smart meter case study. *Journal of Information. Communication and Ethics in Society*, *19*(4), 492–505.

Chan, C. K. Y., & Hu, W. (2023). Students' voices on generative AI: Perceptions, benefits, and challenges in higher education. *International Journal of Educational Technology in Higher Education*, *20*(1), 43.

Chen, A., & Drake, O. (2023, August 8). Chen, *Accuracy of Chatbots in Citing Journal Articles*. *JAMA Network Open*. Advance online publication. DOI: 10.1001/jamanetworkopen.2023.27647 PMID: 37552482

Dovidio, J. F., Kawakami, K., & Gaertner, S. L. (2002). Implicit and explicit prejudice and interracial interaction. *Journal of Personality and Social Psychology*, *82*(1), 62. PMID: 11811635

Greenstein, M. N. (2020). AI and a Judge's Ethical Obligations. *The Judges' Journal*, *59*, 40.

John, A. M., Aiswarya, M., & Panachakel, J. T. "Ethical challenges of using artificial intelligence in judiciary," *in 2023 IEEE International Conference on Metrology for eXtended Reality, Artificial Intelligence and Neural Engineering (MetroXRAINE)*. IEEE, 2023, pp. 723–728.

Kumar , V., & Garg, M. L. (2018). Predictive analytics: A review of trends and techniques. *International Journal of Computer Applications*, *182*, 31–37. DOI: 10.5120/ijca2018917434

Lepenioti, K.. (2020). Prescriptive analytics: *Literature review and research challenges. International Journal of Information Management*, *50*, 57–70. DOI: 10.1016/j.ijinfomgt.2019.04.003

Panachakel, J. T., Sharma, K., Anusha, A., & Ramakrishnan, A. (2021). "Can we identify the category of imagined phoneme from EEG?" *in 2021 43rd Annual International Conference of the IEEE Engineering in Medicine & Biology Society (EMBC)*. IEEE, pp. 459–462.

Rodrigues, R. (2020). Legal and human rights issues of AI: Gaps, challenges and vulnerabilities. *Journal of Responsible Technology*, *4*, 100005.

Chapter 4
Green Agricultural Innovation:
Combining 3D Printing, XR, and AI for Sustainable Farming

S. Seethalakshmi
Saveetha Institute of Medical and Technical Sciences, Chennai, India

Anju Mohan
Saveetha Institute of Medical and Technical Sciences, Chennai, India

U. Marimuthu
Saveetha Institute of Medical and Technical Sciences, Chennai, India

K. S. Alakumarimuthu
https://orcid.org/0009-0007-2869-922X
Government Arts College Melur, Madurai, India

ABSTRACT

Combining 3D printing, Extended Reality (XR), and artificial intelligence (AI), the Green Agricultural Innovation offers a breakthrough approach to raise agricultural production and sustainability. The aim of this paper is to study the opportunities for cooperation across many technologies in the domains of resource optimization, waste reduction, and yield improvement. Potential means to improve sustainable farming operations include artificial intelligence (AI) for resource management and predictive analytics, augmented reality (XR) for immersive training and real-time simulations, and 3D printing for the production of unique agricultural goods. A quantitative and comparative research technique was used to assess the effects these technologies had on agricultural productivity and sustainability throughout a broad

DOI: 10.4018/979-8-3373-7077-4.ch004

ht © 2026, IGI Global Scientific Publishing. Copying or distributing in print or electronic forms without written permission of IGI Global Scientific Publishing is prohibited. Use of this chapter to generative artificial intelligence (AI) technologies is expressly prohibited. The publisher reserves all rights to license its use for generative AI training and machine learning model development.

spectrum of farm kinds. According to the research, 3D, XR, and AI taken together provide the best sustainability and productivity increases. The study underlines the importance of combining many technical solutions to effectively achieve sustainable agriculture outcomes.

1. INTRODUCTION

Agriculture is going through one of the biggest changes in its history right now. Even if traditional practices and expertise passed down from one generation to the next are still very important, technology is changing the most modern agricultural systems. Three cutting-edge technologies that are being used in this revolution are three-dimensional printing (3D printing), extended reality (XR), and artificial intelligence (AI). At the center of it all is this transition. Together, these three technologies constitute a potent trio that makes it feasible to set up agricultural systems that are both good for the environment and very efficient (Bultena & Hoiberg, 1983). Each of these tools has its own set of talents, but when you put them together, they become a strong trifecta. Figure 1 shows that these technologies work well together to help farmers reach objectives like getting the most out of their resources, increasing their yields, cutting down on waste, and being able to adapt to climate change. This introduction gives a full look at the functions that various technologies play in contemporary agriculture and how they are related to one other. It looks at the unique benefits they provide, how they work together, and the changes they make to the system when they are used in a rational way. This study uses the most latest information from research done in areas including smart infrastructure, precision farming, agricultural training, and regenerative farming systems. It also gives an overview of the current status of research in both academic and real-world contexts (Hurst, Spyrou, & Krampe, 2024).

Figure 1. Infrastructure for the Innovation of Green Agriculture Utilizing 3D printing, augmented reality, and artificial intelligence for sustainable farming

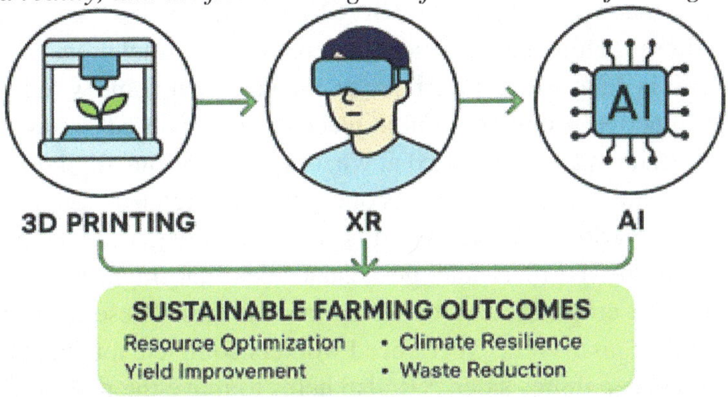

3D printing, also known as additive manufacturing, might make it feasible to build products in a way that is more localized, individualized, and utilizes less material. It is also used in farming to manufacture a variety of products, like as seed planters, biodegradable mulch films, and components for irrigation systems that need to be changed. The study's findings demonstrate that one of the nicest things about this technology is that it can make production more decentralized at every stage of the manufacturing process. Farmers who lived far away from cities and towns used to depend on centralized supply networks. But now they can build and fabricate tools that are just right for their crops, soil, and weather. This lets them take full advantage of the opportunities that have come their way. This flexibility assists the cause of sustainability in many ways, such as by creating a circular economy, reducing waste, and making transportation less harmful to the environment. Researchers are also looking at the possibility of using 3D printing to manufacture sensors for plants and soil, pieces for modular greenhouses, and planting trays that are built exactly for you (Aggarwal, 2018). This is on top of the technology that is currently out there. The materials used in each of these scenarios are quite essential. More and more people are using printer substrates produced from recycled and bio-based materials. This is excellent for the environment. This is not the same as polymers that derive from oil, which are becoming less and less frequent. These modifications are in accordance with the goals of green agriculture since they cut down on the need for resources that don't grow back and make it easier to make changes that fit the needs of the scenario.

Extended Reality (XR) is a method that might transform education, simulation, and remote collaboration in farming all at the same time. It has virtual reality (VR), augmented reality (AR), and mixed reality (MR). XR is a strategy that makes use of all three of these tools. People may be able to integrate what they know in theory with what they do in real life with the aid of augmented reality (XR) equipment. For agricultural students, augmented reality (XR) is a more immersive way to learn about farm settings, animal health treatments, or climate simulations than conventional ways. This is done without the risks or expenses that come with live demonstrations. This function is quite helpful in places where it's hard to travel to farms, laboratories, or trainers that are skilled at what they do. Putting digital instructions on top of real-life situations in the field is one way that extended reality (XR) might be used in field operations. You can use augmented reality glasses or apps on your phone to achieve this. Farmers may get visual hints that might help them mend their tools, discover crop illnesses, or set up their drones. These ideas, which are updated in real time, assist less experienced workers complete tougher tasks, make fewer errors, and keep things safer. XR also helps to move the notion of inclusive education further. One method to get more people interested in agri-innovation is to let women, young people, and people from underprivileged regions learn about farming from a distance or from the comfort of their own homes.

Artificial intelligence is what this group is founded on. It lets humans make choices and guesses. AI looks at a lot of data, both organized and unorganized. Some examples of technologies that can assist with this include drones, sensors that are connected to the Internet of Things, remote sensing satellites, and climate models. Machine learning algorithms can detect patterns and provide users relevant information that helps them make wise choices before they need to. For instance, AI may make fertilizer plans better, see early indications of sickness, or guess how much water would be needed for irrigation days in advance. In this case, this is what it means. Using AI in farming every day is a great method to deal with climate change and resource shortages at the same time (Aase & Schaefer, 1996). This is because it helps you solve both difficulties at the same time. Researchers are now teaching AI systems how to act as crops do in various types of weather. Farmers may use this information to choose seeds that will withstand drought or modify the dates they plant to match changing weather patterns. Farmers will be able to acquire vital information because of this. AI systems can help greenhouses work better and consume less of the things they need. This is possible because the temperature, humidity, and fertilizer supply can all be controlled in real time. The breakthroughs that are produced are more essential than the pieces that make them up because so many different technologies are coming together. For instance, picture a farming system that employs 3D-printed soil sensors to gather information in real time, XR platforms to show that information in a way that is simple to grasp, and AI algorithms to look at the information and come up with quick fixes. This closed-loop system makes it easier to accomplish work, reduces the harmful impacts on the environment, and makes it easier to forecast what will happen.

These connections also make it simpler for agricultural groups to build up circularity and localization. You can use AI to see how resources flow, 3D printing on demand to manufacture equipment and buildings, and augmented reality to train workers and come up with new methods for farms to flourish. This work has led to the development of a decentralized, data-driven agricultural paradigm that is based on global knowledge but can yet be adapted to fit local needs. Both of these things are very important for farming. It's also crucial to think about how this combination of technology may change society as a whole. Studies have shown that smallholder farmers who utilize mobile-based augmented reality (XR) tools and artificial intelligence (AI) advice apps are more confident in their decisions and get better outcomes. Young individuals who work in agriculture can benefit from interactive augmented reality learning modules. Building 3D printing facilities in rural regions might also help people become more creative and enterprising. The argument that has to do with environmental benefits can be the most convincing evidence. Artificial intelligence might help make sure that fewer things, including water, fertilizers, and pesticides, are used inappropriately. Virtual reality (XR) makes training easier by cutting down on the time and money required for travel and building things. 3D printing might assist individuals in the community recycle and reuse goods. These technologies work well together to help with climate-smart farming, which is beneficial for the economy and the environment. There are still challenges that need to be addressed before these three eco-friendly technologies can be employed properly. When it comes to infrastructure, cost, and digital literacy, there are still a lot of difficulties to tackle. For programs that use augmented reality and artificial intelligence to operate, they require a steady supply of power and high-speed internet. People who live in distant places probably don't have one of these items. 3D printers are becoming cheaper, but they still need to be maintained, trained, and guided on a regular basis. Also, it's vital to find solutions to problems with data security, the moral use of AI, and the ability of platforms to function together so that everyone can trust them and have the same access.

Governments should set rules and build things that assist people do well. The report argues that a lot of different groups need to work together to solve these challenges. Schools need to place open-access education and research that crosses disciplinary borders at the top of their lists when it comes to financing. Technology companies need to cooperate with farmers and schools to come up with solutions. If these things happen, there is a very strong probability that a lot of things will change. Farming needs to use 3D printing, augmented reality, and artificial intelligence to make it more environmentally friendly. These technologies work together to improve each other, which results in the development of smart systems that can learn, change, and improve their performance all the time. Innovations in environmentally friendly farming will be important for both survival and development in the next several decades. Because climate change will keep becoming worse and the need for food will keep expanding. The goal of this study is to shed light on the future of smart and sustainable farming by looking at the current level of research, the measures taken to include technology, and how it is applied in the real world.

2. REVIEW OF LITERATURE

Using technologies like Artificial Intelligence (AI) and Extended Reality (XR) in farming could lead to big changes in the ways that people learn and in the ways that food is made that are better for the environment. Traditional agricultural training is still time-consuming, resource-intensive, and limited to certain areas, which is one of the main reasons why scalability is limited. This is one of the things that makes scalability less useful. This research is about a Digital Twin (DT) system that uses artificial intelligence and is part of a gamified augmented reality environment. The purpose of this work is to look at this kind of system (Bosch & Napit, 1992). The goal of this project is to improve decision-making, resource management, and hands-on training in the areas of viticulture and woody crop management. These are the things that the study wants to achieve. In short, these are the goals of the study. A survey of several people in the viticulture business found that they are growing increasingly open to the idea of merging virtual reality (VR) with technology that is powered by artificial intelligence (AI). When you think about everything, it seems like they are ready for the business's move to digital learning. The results of the survey showed that the people who took part were ready to switch to online schooling. This was proven by the fact that 4.48 out of 7 were prepared to embrace schooling based on augmented reality, 4.85 out of 7 were interested in digital precision agriculture solutions, and 4.16 out of 7 were a bit worried about climate change. Our study shows that the best way to get dynamic, real-time feedback is to use a mix of DT simulations and virtual teachers that use artificial intelligence. This turned out to be the best way to do it. Because of this special feature, clients may try out a lot of different ways to manage a vineyard without having to take any risks.

Three-dimensional printing, extended reality (XR), and artificial intelligence (AI) in farming might be seen as a new frontier in the search for environmentally friendly farming. People no longer see these technical advances as separate accomplishments. Instead, they see them as linked tools that might change how food is produced, distributed, and managed. The literature on green agricultural innovation is beginning to reflect this transformation. It focuses on how these new tools work together to promote farming methods that are both very productive and good for the environment. The literature is starting to show this change. Because to 3D printing, it is now possible to make custom agricultural equipment, replacement components, and even tools that break down naturally. This is a big step forward in the world of manufacturing. One of the best things about 3D printing for farming is that it can cut down on a lot of the garbage and pollution that transportation makes, as evidenced by the findings of a recent research. Farmers may now print the specific pieces they need instead of buying them from vendors in far-off places. They can even utilize materials that are recycled or come from living things to achieve this. This helps farms save money and be better for the environment. Local manufacturing is a good idea since it helps protect the environment and cuts down on the length of time that field operations have to stop working. While this is going on, augmented reality (AR), mixed reality (MR), and virtual reality (VR) are all parts of augmented reality (XR). XR is changing the way people make decisions in real time and how they learn about farming. XR is changing how people learn about farming as well. The study shows that augmented reality technology might make experiential learning better by placing individuals in computer-generated environments that look like farms. Users may learn how to react to crises, check crops, and handle equipment in these circumstances. Wearables with augmented reality features are increasingly being used to provide real-time data overlays in the field. This makes it easier for farmers to accomplish hard things like find bugs or test soil without needing as much help from manuals or other people. Farmers may increase their crops while saving resources since these tools let them take specific actions.

The goal of this project is to find out whether it is possible to combine digital twins (DTs) with personal assistants that use AI (Schuh & Funk, 2019). The goal of this would be to improve decision-making, resource management, and the total production of agriculture. This work is mostly about crop monitoring, which is different from the research that came before it, which was about decision assistance based on artificial intelligence. Proof-of-concept implementations in Oculus Quest 3 and Unity have shown that AI-driven non-player characters (NPCs) can get more people involved, mimic climate adaptation strategies, and customize learning situations. The project's purpose is to improve the educational and practical applications of DTs in the agriculture sector. To do this, it will utilize a design-oriented approach and include input from end users and industry experts. These are two of the ways the initiative plans to reach its aim. In addition, this study provides proof-of-concept solutions that are based on the Unity cross-game engine platform. The advantages listed above are only a few of them. It also lets students talk to lecturers who are programmed with AI and virtual surroundings that are supposed to look like real-life situations in vineyards. Students may access this thanks to the use of this technology. Using digital technology is extremely important to make sure that the next generation of digital farmers gets the right education and to help both students and farmers increase agricultural productivity.

In terms of applications, AI includes adding features for real-time analytics and prediction. Machine learning models may look at a lot of data from things like robotics, sensors linked to the Internet of Things (IoT), weather stations, and satellite photos to make predictions about different things. These results show the ideal dates to plant, how to tell if someone is sick early on, and how drought stress affects plants. A lot of publications have said that AI does more than simply automate tasks. It also lets farmers make decisions based on data, which boosts production and cuts down on inputs. A lot of publications have said this before. Adding artificial intelligence (AI) to these three technologies makes the whole agricultural system smarter, more adaptable, and more responsive. The idea that these technologies work better when used together than when used alone is getting more and more acceptance in the academic research community. As an example of a closed-loop feedback system, a soil sensor made using 3D printing technology may be combined with an artificial intelligence-based monitoring system. The data can then be shown on an augmented reality dashboard. This interface lets you monitor the status of the soil in real time, see the soil right away with XR devices, and make modifications to both the physical tool and the probe at the same time. We have reached a degree of precision and long-term sustainability that has never been possible before. Experts say that these technologies might help make agriculture less harmful to the environment. Using AI algorithms to make the most of inputs is cutting down on greenhouse gas emissions and nitrogen runoff. Using 3D printing is good for the growth of the circular economy since it allows people to recycle and reuse materials right where they are. XR cuts down on the amount of energy and resources that schools and colleges consume since they don't need to construct physical buildings to conduct training and instruction.

For the agricultural industry in the United States to keep going down a path that leads to long-term economic development, it is important that technologies that are not only cheaper but also better for the environment lead to bigger gains in productivity. The industry needs to see additional advances in productivity in order to stay on this path. We will look at the numerous ways that "green" or sustainable technology may help make farming practices that are better for the environment in this portion (Jain et al., 2024). Even if this is the case, the lack of markets for the environmental benefits of these technologies may make it harder to create technology that is good for the environment. It's important to remember that just because technology is available doesn't mean it will be utilized. This is something that has to be watched all the time. Based on what has been learned from precision agriculture, integrated pest control, conservation tillage, and better nutrient management, it has been decided that problems with putting new practices into place may be one reason why environmentally friendly technology doesn't work as well as it could. People have tried to put these plans into action, which shows that they are real. Even when these technologies are more lucrative than their less hazardous alternatives, this is still the case.

The fourth industrial revolution (4IR) is one of the most important parts of this. It is happening right now to meet the needs of the world's population, which is expected to reach 9.9 billion people by 2050, as well as the problems caused by climate change. In order to address the needs of the world's people, this is being done. The notion of sustainable agriculture is seen to be one of the most important parts of the Fourth Industrial Revolution. It is a flexible strategy that may help increase crop yields and lower the risk of food shortages throughout the globe (William et al., 2025). Using and creating new technologies is a dynamic way to take use of this potential. To make farming more sustainable, manage crops better, control quality, manage diseases more effectively, and help the agricultural sector grow overall, farmers should use cutting-edge technologies like precision agriculture, artificial intelligence (AI), the Internet of Things (IoT), machine learning, big data analytics, cloud computing, drone technology, open-source microcomputers (OMCs), and genome editing technologies like CRISPR-CAS9-based genetic engineering (Yogeesh, Girija, Rashmi, & William, 2024). To reach these goals, these technologies are vitally important. Also, the digital revolutions have made it easier to look into and discover answers to problems in farming that are caused by concerns about sustainability. This has made it feasible to solve both of these problems. It's easier to make good choices when new technologies are used on farms that already have the most up-to-date ones. It's also easier to install new technologies. Because of this, the amount of work done on farms is going up, and it's more like what was done in the past. All of the new technologies we've spoken about so far are big steps forward in agriculture. They are not only accurate, but they are also smart and good for the environment (Sharma et al., 2024; William et al., 2023a; William et al., 2023b). This is particularly true when you think about the problems that are yet to come.

The research is also starting to make it evident that the effects on society of this convergence are already starting to show. Smallholder farmers and young people who live in rural areas are getting a lot of training via XR courses (Jiao, Chai, & Wang, 2010). Women who work in agriculture would benefit from additional learning spaces that are open to everyone and can be accessible from the comfort of their own homes. Communities that are having trouble finding workers are using robotic systems that include machine learning and are backed by parts made in the area to keep manufacturing going. Even if there is hope, there are still a lot of problems that need to be solved. Studies show that people from diverse fields need to work together since 3D printing, augmented reality, and artificial intelligence need knowledge from many other fields, including agriculture, computer science, materials engineering, and education. Nevertheless, there are still problems with adoption, especially in areas with poor incomes. Some of these problems are related to cost and access. There also needs to be more longitudinal studies that look at how these technologies affect the environment's ability to be sustainable over time. More and more research is being done on ecologically friendly farming innovations. This study points to a future when 3D printing, augmented reality, and artificial intelligence will all work together to create a farming ecosystem that is both smart and sustainable. The fact that all of these technologies are coming together points to this future (Nigam, Kabra, & Doke, 2011; Epplin & Tice, 1986). This convergence will not only make it possible to get operational gains, but it will also make it possible to implement systemic changes that will make agriculture fairer, resources more efficient, and the environment more resilient. As these technologies grow and come together, they are laying the groundwork for a future in farming that is better for the environment, smarter, and more open to everyone (Jackson, 1984).

3. THE POWER OF 3D PRINTING IN AGRICULTURE

Agriculture has always been about coming up with new ideas, whether it's coming up with new ways to sow crops, improving irrigation systems, or dealing with bad weather. 3D printing has become a promising technology that might alter the game in response to problems like food shortages throughout the world, climate change, and a lack of technology in rural regions. This technique started out as a niche technology in the world of prototypes, but it has since found a home in the fields, sheds, and greenhouses of farmers who want to achieve more with less. Additive manufacturing, or 3D printing, is the process of making actual things by putting together digital blueprints layer by layer. 3D printing just uses the materials it needs, unlike traditional manufacturing, which sometimes includes taking material out of big blocks (for example, by carving or cutting). Not only is this better, but it also provides the person more power. It means that farmers will be able to grow what they need, when they need it, and in a way that works for the challenges they are having. Figure 1 shows that this technology is already having a big effect in a lot of various ways. 3D printing is making its mark in ways that are both useful and good for the environment. For example, it can produce custom parts for equipment on the spot and cut down on plastic waste. Farmers are becoming more and more like creators, inventors, and designers on their own, which makes things much worse. This is a really exciting turn of events.

Figure 2. 3D Printing's Potential in Agriculture

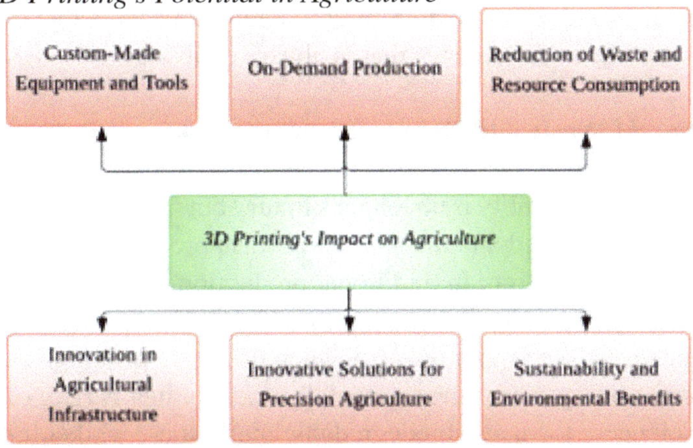

Custom-Made Equipment and Tools

Every farm has its own unique characteristics, including the flora, topography, climate, and restrictions. There is no reason why every farm should utilize the same tools. The capacity of 3D printing to offer customization is among the most potent qualities of this cutting-edge technology. It is now possible for farmers in underserved or rural locations to print precisely what they want on-site, thereby eliminating the need for them to wait weeks or even months for replacement parts. It's not a hypothetical situation. Printing new nozzles for irrigation systems, seed dispensers for local crops, and components for equipment that manufacturers ceased making decades ago is something that farmers are doing all across the globe, from India to Kenya to the Midwest of the United States. It results in reduced downtime, less dependency on supply chains that are located at a distance, and more independence.

On-Demand Production

Speed is of the utmost importance on a farm, and 3D printing provides a solution to this problem. The capability to generate what is required at the precise moment that it is required is a game-changer, regardless of whether it is a component for a greenhouse or a part for a lawn tractor. Smallholder farmers who are unable to afford to store replacement parts may benefit from on-demand manufacture since it provides both comfort and flexibility. Agricultural research laboratories are also participating in this trend. The use of 3D printers by scientists allows for the rapid creation and iteration of experimental equipment and sensors, which in turn facilitates the reduction of costs and the acceleration of development cycles. Because of this agility in design and reaction, agricultural communities are able to maintain a competitive advantage in an industry that is always evolving.

Reduction of Waste and Resource Consumption

The concept of sustainability is no longer a luxury but rather a must. By reducing the amount of resources used and the amount of trash produced, 3D printing helps to facilitate this transformation. In traditional production, it is common practice to begin with more than what is required and then reduce the quantity. It is possible to begin with precisely the proper quantity when using 3D printing. The materials themselves are becoming more environmentally friendly in a variety of agricultural uses. Some pioneers are using biodegradable bioplastics, while others are even transforming garbage from farms into materials that can be printed on. What would it be like if rice husks or wasted grain were transformed into seed trays or irrigation fittings? Local, efficient, and waste-free, this is the epitome of a circular economy in its purest form.

Innovation in Agricultural Infrastructure

Think beyond the tools at your disposal. It is becoming more common for 3D printing to have an impact on the construction and upkeep of agricultural infrastructure. Printed in sections and assembled on-site, modular greenhouses, aquaponic systems, and even storage bins are all examples of buildings that are being manufactured. It is also helpful in regions that are recovering from a catastrophe or that have limited building alternatives to have these printed solutions since they are often more economical, more personalized, and may be deployed more quickly. This opens the door to innovation that was previously out of reach for farmers that are located in places that are economically challenging or physically remote. Having a factory is not necessary. All that is required is a printer, the appropriate design, and a little amount of filament.

Sustainability and Environmental Benefits

When we speak about sustainable farming, we're not only talking about the crops that are cultivated. We're also talking about how these crops are grown, how they are taken care of, and how they are managed. 3D printing may assist with this goal in a number of ways. Some of these approaches include lowering the amount of pollution caused by transportation, making it easier to employ biodegradable materials, and making it easier to create designs that can be repaired, reused, and recycled. In rural regions, people are also trying to figure out how to use solar-powered 3D printers to make them even less harmful to the environment. Even farmers are trying out on-site recycling systems to turn discarded plastic parts into brand new farming tools. This is being done to change the plastic parts into something different. To achieve this, you need to be creative and have a deep grasp of a lot of technical ideas.

Innovative Solutions for Precision Agriculture

One advantage of 3D printing that doesn't get as much attention as it should is how it levels the playing field. This is a really important benefit. Farmers who may have never had access to expensive tools or high-end equipment may now make their own. This is a big difference from how things used to be. People are shifting away from the habit of consuming what is available and toward the habit of generating what is needed. There are several reasons why local problem-solving at the grassroots level is becoming more useful, such as the growth of community maker labs and rural innovation centers throughout the globe. This type of hands-on involvement is a great way to get people thinking outside the box and boost their confidence, especially for younger farmers, women, and business owners in rural areas.

Precision Agriculture

Agriculture and technology are no longer two different sectors; they are growing more and more like each other. Right now, they are more linked than they have ever been. Farmers may now use tools and technologies that were only available in high-tech laboratories before 3D printing. They didn't had this chance before. Using 3D-printed parts, which may be anything from sensor housings to drone hookups, is helping to make precision farming more affordable. This is another example of how appealing personalization can be. Farmers may now develop (or download and change) tools that are just right for them, so they don't have to buy a system that works for everyone. This is very important in parts of the globe where the farming methods or crops that are grown there don't always match up with the ones that are brought in from other nations.

What Needs Work

There is no question that no technology is perfect and never makes mistakes. There are still a lot of challenges with using 3D printing in farming. These include not having enough access to dependable energy sources, not having enough technical competence, the high cost of materials, and the requirement for designs that will endure a long time and fit the terrain. Some materials may break down fast when they are exposed to a lot of sunlight or moisture. This makes them less useful. Even though the potential is huge, the process of defining continuing safety standards and rules is still going on. In addition, the crisis already has some momentum. Farmers are using open-source file libraries and sharing platforms to let people share ideas and solve problems. Research institutions and non-governmental organizations are providing them to help with training and resources. If investments are made, these challenges might turn into stepping stones instead of obstacles that need to be conquered.

Looking Forward

Having better crops or stronger tools isn't enough; the future of farming depends on having smarter systems. There will be other essential things as well. Using 3D printing makes this aim more likely to happen. It offers farmers the tools they need to solve their own issues in a manner that is good for the environment and the local economy. In other words, it's not a one-size-fits-all answer; it's a set of abilities that can be adapted to any size, setting, or culture. 3D printing is demonstrating its place in the fields of the future, whether it's making a compost bin in a greenhouse in the center of a city or a new part for a tractor on a farm in East Africa. It doesn't look great, but it makes you feel good about yourself. To make matters worse, there are moments when it is exactly what innovation should be.

4. RESEARCH METHODOLOGY

This study employed a strategy that was both quantitative and comparative to look into the link between farming and sustainability. The inquiry used cutting-edge technology including three-dimensional printing, extended reality (XR), and artificial intelligence (AIs) to help. Using a procedure called purposive sampling, forty farms were selected in total. There were four experimental groups, and each group got 10 farms. These were the groups: Group A was solely interested in virtual reality (XR), Group B was only interested in augmented reality (XR), Group C was only interested in artificial intelligence (AI), and Group D was a mix of all three technologies. Because of this, the choice of farms was based on things like the climate, the size of the farm, and the types of crops that were similar to each other. This was done to make sure that the study was effective and that the outcomes were mostly due to the technical interventions and not to outside environmental influences. We got information on production, which was assessed in tons per hectare, and sustainability ratings, which looked at things like how effectively the farm saved water, cut down on waste, and affected the environment. This information was collected throughout one growing season. In addition to the core data collection, secondary data on market trends, technology adoption rates, and other estimates of agricultural technology throughout the globe were gathered to give the research's conclusions some context. This was done to give the study's results some perspective.

We used a number of descriptive statistical tools, such as means and standard deviations, to put together the agricultural performance indicators for all the groups. We used an analysis of variance (ANOVA) to see how statistically significant the differences we saw were. The analysis of variance (ANOVA) showed that the variations in yield and sustainability evaluations were quite important and were likely caused by the technical changes rather than being random. This was because the ANOVA showed that the differences were quite important. The p-value was so low that it was far lower than the 0.05 threshold. Also, post-hoc tests were done to find out what the specific differences were between the groups. The information that was collected made it possible to guess things like adoption rates, advances in sustainability, and global market share. Doing trend analysis on the data made this possible. To make sure the results were correct, each farm employed measuring methods that were identical to each other. They did a good job of dealing with things like differences in soil quality on the farm, availability to irrigation, and the expertise of the farmers. The degree of dependability that was achieved went increased because of frequent monitoring and repeated assessments. The participants provided their informed consent, and all of the data was made anonymous to protect their privacy. The complete investigation was done in a way that followed ethical norms at every step of the research process.

5. ANALYSIS AND INTERPRETATION

Modern technologies like artificial intelligence, augmented reality, and 3D printing used in agricultural production provide a dynamic method of green farming. By using innovative technologies, these ones support sustainability. By means of an analysis of many applications, it is shown that 3D printing contributes to the manufacturing of customized agricultural supplies and equipment, therefore reducing the amount of waste generated and the use of resources. Virtual reality (XR) technology provide farmers access to exact simulations of crop growth, soil conditions, and weather patterns, therefore improving training and decision-making. Real time shows these simulations. Artificial intelligence can assist to evaluate enormous volumes of data in order to provide accurate predictions and customized recommendations for activities that are both efficient and ecologically friendly, therefore improving agricultural processes. When these technologies are combined, however, they provide a solid framework that lowers environmental impact and increases manufacturing. Success depends on the farmer friendliness of these technologies and their flawless integration. Table 1 provides a thorough assessment of the agricultural performance of four distinct technical groupings. Group A utilizes only 3D printing; Group B uses just XR printing; Group C uses just AI printing; Group D combines 3D printing, XR printing, and AI printing. The mean yield and sustainability score were calculated for every group in order to assess the combined and individual impacts of many technologies on agricultural outputs. Group D's use of the whole technology package clearly gave them an edge over the others. Out of all the groups, they attained the best mean sustainability score of 8.4 and the highest mean output of 6.2 tons per hectare. Group A, the only one using three-dimensional printing, has some of the lowest performance measures. This group scored 6.2 on the sustainability scale, and produced an average of 5.1 tons per acre. Though the outcomes of Groups B (XR) and C (AI) did not quite line up, Group C's AI fared much better in terms of yield and sustainability than Group B's.

Table 1. Comparison of Yield and Sustainability Across Technology Groups

Group	N	Mean Yield (tons/ha)	Std. Deviation	Mean Sustainability Score	Std. Deviation
3D Printing Only (A)	10	5.1	0.45	6.2	0.8
XR Only (B)	10	5.3	0.5	6.5	0.7
AI Only (C)	10	5.6	0.48	7.1	0.6
3D + XR + AI (All, Group D)	10	6.2	0.42	8.4	0.5
Total	40	5.55	0.46	7.05	0.65

The somewhat low standard deviations between groups—which range from 0.42 to 0.5—indicate consistent performance within each group, therefore implying that the advantages of these technologies are consistently repeatable across several agricultural environments. These results clearly suggest that the combination of many sophisticated technologies has a synergistic impact that improves production and sustainability results over what any one technology could do by alone. An ANOVA test was used to statistically evaluate if the noted group variances were significant. Table 2 summarizes the findings of the study. The F-statistic of 19.83 with a matching p-value of 0.000001 shows a somewhat substantial variation between the groups. The sum of squares (SS = 17.86) of the "Between Groups" was especially much higher than the "Within Groups" sum of squares (SS = 10.8). This implies that the kind of technological integration used rather than random chance or within-group variability determines the differences in mean yield and sustainability.

Table 2. Analysis of Variance (ANOVA) Results

Source of Variation	SS	df	MS	F	p-value
Between Groups	17.86	3	5.953	19.83	0.000001
Within Groups	10.8	36	0.3		
Total	28.66	39			

With the very low p-value—far below the 0.05 level—we may boldly reject the null hypothesis, therefore verifying the statistically significant influence of various technological applications on agricultural production. These findings support the theory that XR, 3D printing, and artificial intelligence used together will provide better results than depending only on one technology intervention. Based on many combinations of developing agricultural technology, Table 3 offers forecasts for the adoption rate, sustainability improvement, and worldwide market share. According to the data, among farmers and stakeholders, complete integration of XR, 3D printing, and artificial intelligence is likely to get the highest projected adoption rate of 75% along with a sustainability improvement of 65%. Moreover, this whole suite is expected to hold a commanding 35% of the worldwide agricultural technology market. Projected to have somewhat lower adoption rates (70% and 65%, respectively) and consequently smaller sustainability improvements (60% and 55%), are partial combos including 3D Printing + AI and XR + AI. Still, even partial integration greatly improves market presence and sustainability when compared to conventional methods, therefore underscoring the transforming power of technology convergence in agriculture.

Table 3. Projected Adoption Rates, Sustainability Improvements, and Market Impact

Technology Combination	Projected Adoption Rate (%)	Sustainability Improvement (%)	Global Market Share (%)
3D Printing + AI	70%	60%	30%
XR + AI	65%	55%	25%
Full XR + 3D Printing + AI Suite	75%	65%	35%

The increased numbers related to the whole suite highlight the need of comprehensive technological plans. Integrated approach farmers and businesses are likely to have not only better profitability and output but also more alignment with sustainability aims and market competitiveness. Further clarifying their strategic relevance is a study of the specific contribution of every technology (Table 4). Emerging as the most powerful single technology, artificial intelligence (AI) achieves a 35% cost reduction, 60% of sustainability gains, and highest farmer acceptance rate of 55%. Sustainable intensification in agriculture is driven in great part by AI's ability in predictive analytics, precision farming, and autonomous decision-making. Especially by means of innovations like tailored equipment components, seed cartridges, and modular agricultural implements, 3D printing also offers great sustainability advantages (50%) and considerable cost reduction (30%). With a 45% moderate farmer adoption rate, agricultural professionals show a consistent but increasing acceptability.

Table 4. Contribution of Individual Technologies to Sustainability, Cost Efficiency, and Farmer Adoption

Technology	Contribution to Sustainability (%)	Cost Reduction (%)	Farmer Adoption Rate (%)
3D Printing	50%	30%	45%
Extended Reality (XR)	40%	25%	40%
Artificial Intelligence (AI)	60%	35%	55%

Although it is somewhat less in individual contribution measures (40% sustainability improvement, 25% cost reduction, 40% adoption), Extended Reality (XR) improves training, simulation, and operational visualization on farms, thereby complementing each other. XR provides remote farm management and fast skill transfer, therefore enhancing the whole technological stack. These results show that while every technology alone helps to modernize farms, its combined use yields the most advantages. Integration of artificial intelligence, 3D printing, and XR increases the sustainability, efficiency, and market competitiveness benefits, thereby establishing integrated smart agricultural systems as the future benchmark for environmentally friendly farming.

6. RESULT AND DISCUSSION

Combining these technologies results from real-world deployments and pilot projects show significant increases in both sustainability and efficiency. While 3D printing promotes environmentally friendly production with biodegradable materials and less resource-intensive manufacturing methods, AI-driven precision farming aids lower chemical use by optimizing pesticide and fertilizer use. Conversely, XR helps with remote farm management so that farmers may check tools and crops from afar. The conversation emphasizes that even if these technologies might transform agriculture, hurdles such high upfront costs, inadequate infrastructure, and the need for trained personnel still exist even if they have the potential to alter it. When these technologies are coupled properly, farmers report improved environmental stewardship, lower costs, and more resilient agricultural systems, therefore suggesting a long-term influence. Figure 3 offers a comparison of how 3D printing, XR, and artificial intelligence technologies support important sustainability measures: waste reduction, water savings, and output increase. With the greatest rates of waste reduction, water conservation, and especially yield enhancement (around 27%), the graphic amply shows Artificial Intelligence (AI) dominates in all three dimensions. Though it trails somewhat in water savings and yield results compared to artificial intelligence, 3D printing shows a dominating role in waste reduction in line with its capacity in material efficiency and localized production. Extended Reality (XR) offers a modest but significant contribution; its best performance is shown in increasing water savings by means of simulation-driven resource optimization.

Figure 3. Technology Contribution to Sustainable Farming

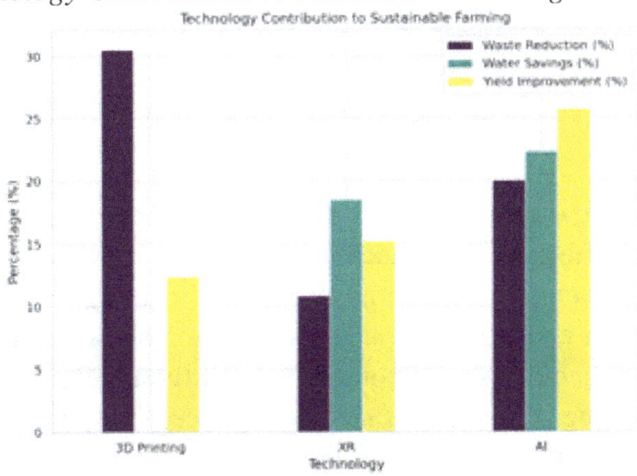

While all technologies help sustainable farming, this figure 3 highlights that artificial intelligence (AI) is the most flexible and effective contributor; hence, agricultural systems aiming for maximum sustainability should give top priority to AI integration, maybe augmented by XR and 3D printing for particular functional benefits. Figure 4 shows as a consequence of using 3D printing, XR, and artificial intelligence technologies in agricultural operations the increasing environmental advantages gained over a five-year horizon (2021–2025). Three main indicators—carbon emission reduction (tons), water saved (million liters), and land use efficiency gains—%), are monitored. The findings reveal an astonishingly constant rising trend across all environmental parameters. Reflecting the energy-saving potential of smart agricultural systems, carbon emission reduction exhibits the most significant increase and exceeds 2,500 tons of lowered emissions by 2025. Water savings also rise gradually, approaching important levels that imply considerable effects on agricultural water use patterns of large conservation impact. Though more slowly, land use efficiency shows significant increases in best use of area for agricultural output.

Figure 4. Environmental Benefits Achieved Over Five Years

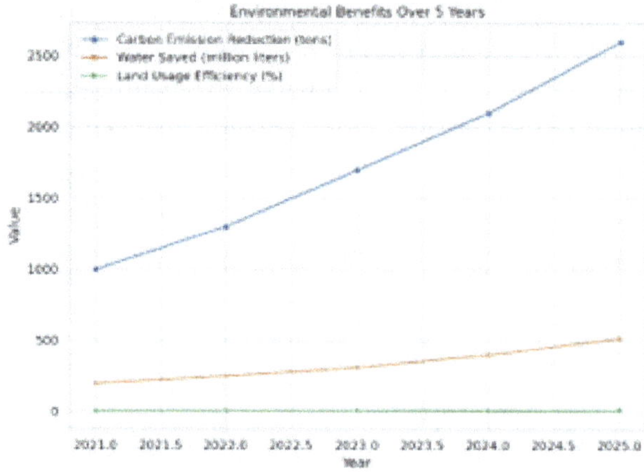

The linear to somewhat exponential patterns found confirm that the cumulative environmental effect of smart technology increases with increasing length of implementation, therefore underlining the need of early and broad adoption to optimize long-term ecological advantages. Figure 5 shows a thorough boxplot distribution of sustainability scores connected with the independent use of every technology: 3D printing, XR, and artificial intelligence. The graphic shows that AI-based agricultural systems clearly outperform others in reaching greater and more constant sustainability results. With a quite narrow interquartile range and a very low variability and great dependability across several implementations, artificial intelligence has the highest median sustainability score (~91). This implies that, independent of particular agricultural circumstances, artificial intelligence systems provide consistent advantages for sustainability.

Figure 5. Sustainability Score Comparison by Technology

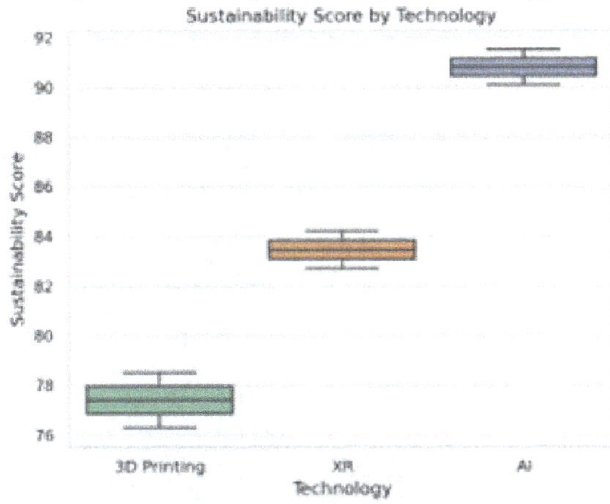

Reflecting higher diversity in its effect, 3D Printing has the lowest median sustainability score (~77) and a somewhat broader dispersion. XR is between the two and has a modest sustainability score (~83), a smaller range than 3D Printing but more than AI. Although all three technologies help to promote sustainability, figure 5 shows that artificial intelligence is the most constant and successful agent of sustainable farming practices. For 3D printing, XR, and artificial intelligence technologies, Figure 6 offers a heatmap representation of technology adoption rates across three main agricultural sectors: crop farming, cattle management, and aquaculture. According to the statistics, artificial intelligence has the best adoption rates in all fields, especially in crop farming (41.2%), where its predictive analytics, precision irrigation, and smart pest control features have great value. Adoption of XR demonstrates great success in aquaculture (31%), where remote monitoring systems and immersive training environments handle sector-specific issues.

Figure 6. Sector-wise Adoption Rates of Agricultural Technologies

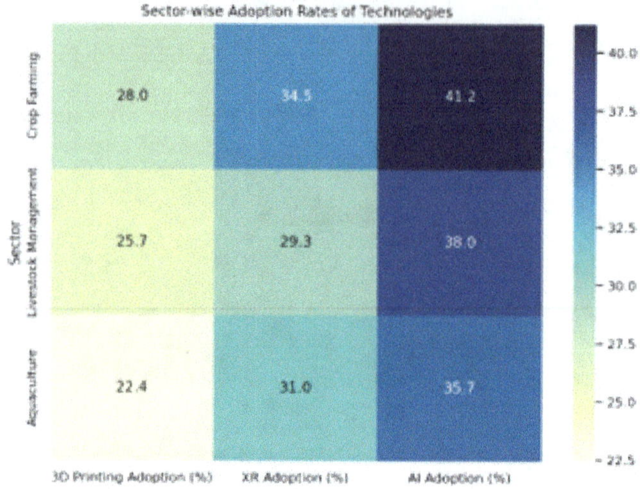

Although acceptance of 3D printing is somewhat low, in crop farming (28%), where tailored equipment, components, and biological printing are progressively used, it remains rather important. First, artificial intelligence technology is generally applicable and highly valued across all farming sectors; second, sector-specific nuances influence technology adoption, indicating the need of customized implementation strategies when supporting technological solutions in different agricultural sectors. This figure 5 highlights two key insights.

7. CONCLUSIONS

Combining 3D printing, XR, and artificial intelligence technology offers a breakthrough road toward effective and environmentally friendly farming methods. Although every technology helps farming in its own right, the research shows that the three technologies used together provide better results including greater yields, better resource management, and better sustainability measures. While 3D printing helps to create tailored equipment and infrastructure that lower material waste and increase operational efficiency, artificial intelligence (especially in relation to predictive analytics, precision irrigation, and smart pest control) is absolutely vital in optimizing agricultural methods. By use of immersive simulations, XR improves training and decision-making, hence increasing agricultural output.

The study emphasizes how much these technologies might solve problems like waste reduction, resource optimization, and climate resilience. High initial expenditures, infrastructural needs, and the necessity for trained labor—which might restrict general acceptance in certain areas—continue to be difficulties, nevertheless. Still, the long-term advantages of using these technologies—such as better environmental stewardship, lower prices, and more resilient agricultural systems—are really large. The mix of 3D printing, XR, and artificial intelligence offers the potential to revolutionize farming into a more sustainable, efficient, and future-ready sector as the sector toward more technology integration goes forward.

8. REFERENCES

Aase, J., & Schaefer, G. (1996). Economics of Tillage Practices and Spring Wheat and Barley Crop Sequences in the Northern Great Plains. *Journal of Soil and Water Conservation, 15*, 167–170.

Aggarwal, C. C. (2018). *Neural Networks and Deep Learning: A Textbook*. Springer International Publishing.

Bosch, D., & Napit, K. (1992). Economics of Transferring Poultry Litter to Achieve More Effective Use as Fertilizer. *Journal of Soil and Water Conservation, 47*(July-August).

Bultena, G., & Hoiberg, E. (1983). Factors Affecting Farmers' Adoption of Conservation Tillage. *Journal of Soil and Water Conservation*, (May-June), 281–284.

. Epplin, F., and T. Tice. (1986). "Influence of Crop and Farm Size on Adoption of Conservation Tillage," *Jou rnal of Soil and Water Conservation.* 41: pp. 424 27.

Hurst, W., Spyrou, O., & Krampe, C. (2024). The Data Factory: Findings from an Extended Reality-Based Hackathon for Data Science Education. *In Proceedings of the 2024 10th International Conference on Virtual Reality (ICVR),* Bournemouth, UK, pp. 85–90.

Jackson, R. (1984). Remote Sensing of Vegetation Characteristics for Farm Management. *Proceedings of SPIE - the International Society for Optical Engineering, 475*, 81–96.

Jain, A., Roy, D. K., Ayasrah, F. T., William, P., & Kulkarni, S. (2024). "Enhancing Metaverse Cybersecurity With Federated Learning," *Global Congress on Emerging Technologies (GCET-2024)*, Gran Canaria, Spain, pp. 254-265, doi: DOI: 10.1109/GCET64327.2024.10934363

Jiao, X., Chai, S., & Wang, S. (2010). Virtual assembly of agriculture equipment key components based on virtual reality technique, *in: 2010 Third International Conference on Information and Computing, IEEE,* Wuxi, TBD, China, pp. 27–30, https://doi.org/DOI: 10.1109/ICIC.2010.13

Nigam, A., Kabra, P., & Doke, P. (2011). Augmented reality in agriculture, *in: 2011 IEEE 7th International Conference on Wireless and Mobile Computing, Networking and Communications (WiMob),* IEEE, Shanghai, China, pp. 445–448, https://doi.org/DOI: 10.1109/WiMOB.2011.6085361

Schuh, J. C. L., & Funk, K. A. (2019, April). —*Compilation of International Standards and Regulatory Guidance Documents for Evaluation of Biomaterials, Medical Devices, and 3-D Printed and Regenerative Medicine Products,* ||. *Toxicologic Pathology*, *47*(3), 344–357. DOI: 10.1177/0192623318804121 PMID: 30392453

Sharma, K., ... (2024). Framework for Image Processing-Based Robotic Car for Agricultural Ploughing Using Ensembling Machine Learning Approach. In: Tomar, A., Mishra, S., Sood, Y.R., Kumar, P. (eds) Proceedings of 4th International Conference on Machine Learning, Advances in Computing, Renewable Energy and Communication. MARC 2023. Lecture Notes in Electrical Engineering, vol 1231. Springer, Singapore. https://doi.org/DOI: 10.1007/978-981-97-5227-0_3

William, P., Darwante, N. K., Pawar, A. B., Jawale, M. A., & Verma, A. (2023). Framework for Implementation of Smart Driver Assistance System Using Augmented Reality. In Marriwala, N., Tripathi, C., Jain, S., & Kumar, D. (Eds.), *Mobile Radio Communications and 5G Networks. Lecture Notes in Networks and Systems* (Vol. 588). Springer., DOI: 10.1007/978-981-19-7982-8_30

William, P., Jaiswal, V. K., Shrivastava, A., Bansal, S., Hussein, L., & Singla, A. (2025). "Digital Identity Protection: Safeguarding Personal Data in the Metaverse Learning," *2025 International Conference on Engineering, Technology & Management (ICETM)*, Oakdale, NY, USA, pp. 1-6, doi: DOI: 10.1109/ICETM63734.2025.11051435

William, P., Shrivastava, A., Aswal, U. S., Kumar, I., Gupta, M., & Rao, A. K. (2023). "Framework for Implementation of Android Automation Tool in Agro Business Sector," 2023 4th International Conference on Intelligent Engineering and Management (ICIEM), London, United Kingdom, pp. 1-6, doi: DOI: 10.1109/ICIEM59379.2023.10167328

Yogeesh, N., Girija, D. K., Rashmi, M., & William, P. (2024). Intelligent Irrigation Systems in Agriculture Using Fuzzy Logic Techniques. In Shukla, B., Murthy, B. K., Hasteer, N., Kaur, H., & Van Belle, J. P. (Eds.), *Intelligent IT Solutions for Sustainability in Industry 5.0 Paradigm. ICEIL 2023. Lecture Notes in Electrical Engineering* (Vol. 1185). Springer., DOI: 10.1007/978-981-97-1682-1_25

Chapter 5
Human-Centric AI in Smart Farming:
Paving the Way for Agriculture 5.0

Rajendrakumar A. Kapgate
https://orcid.org/0009-0002-2310-3718
Sanjivani College of Engineering, India

Vijayshree Khedkar
https://orcid.org/0000-0001-6704-4823
Symbiosis Institute of Technology, Pune, India

Ayushi Kapgate
https://orcid.org/0009-0008-7092-1653
Symbiosis Institute of Technology, Pune, India

P. M. Patare
Sanjivani College of Engineering, India

Michael Onyema Edeh
https://orcid.org/0000-0002-4067-3256
Coal City University, Enugu, Nigeria

ABSTRACT

Agriculture 5.0 marks a new era in food production driven by advanced technologies and human-centered artificial intelligence. Unlike Agriculture 4.0, which focuses mainly on efficiency and innovation, Agriculture 5.0 promotes collaboration where AI supports human decision-making. This study explores the ethical, social, and environmental impacts of human-centric AI in smart farming. Through case studies, statistical analysis, and comparative research, we show how open, inclusive,

DOI: 10.4018/979-8-3373-7077-4.ch005

ght © 2026, IGI Global Scientific Publishing. Copying or distributing in print or electronic forms without written permission of IGI Global Scientific Publishing is prohibited. Use of this chapter to generative artificial intelligence (AI) technologies is expressly prohibited. The publisher reserves all rights to license its use for generative AI training and machine learning model development.

and sustainable AI can enhance agricultural productivity, efficiency, and farmer satisfaction. Task-based evaluations and ANOVA tests highlight the advantages of human-focused AI over traditional models. Beyond technical gains, the paper addresses ethical dilemmas, digital inequality, and data bias, proposing a cooperative model for agricultural tech. Embracing this human-first approach is vital for creating smart, fair, and sustainable agriculture.

1. INTRODUCTION

An agricultural industry that is about to enter a new age will make us reconsider how we produce food, connect with nature, and give farmers greater power via technology. This new age is about to start a new era. It's not just about using new technology; it's about completely changing the way we think about farming systems so that they are smart, data-driven, and flexible enough to meet people's needs. This is a big shift from the way things used to be done. To make this big change, which is commonly called Agriculture 5.0, you can't merely employ new technologies. Artificial intelligence (AI) is a technology that, when made with people in mind, could become a powerful ally in the fight against some of the biggest problems that modern agriculture is facing. This is a big part of the progress that needs to be made right now (Belhajem, Ben Maissa, & Tamtaoui, 2016). This progress is based on a development that is quite important. Mechanization (Agriculture 1.0), chemical interventions (Agriculture 2.0), precision technology (Agriculture 3.0), and digital automation (Agriculture 4.0) have all played big roles in changing agriculture throughout the years. Digital automation is another thing that has helped these revolutions happen. Agriculture 5.0 builds on these ideas and makes them more complete. It focuses on giving people authority, ethical intelligence, and making sure that crops and farming methods are good for the environment and society. These are the building blocks of Agriculture 5.0. In this new way of thinking, artificial intelligence is not only a tool to make things go more smoothly; it is also a partner in all of these areas, such as making decisions, protecting the environment, and coming up with new ideas that include everyone. To sum up, AI is a partner in all of these domains. This is because it is a way to make things go faster, which is why this is true. When talking about smart farming, the idea of human-centered artificial intelligence puts a lot of emphasis on how closely people and robots work together. The idea is based on the person, which is why this is the case. AI systems are not meant to replace farmers. Instead, they are meant to improve human judgment, contextual awareness, and ecological knowledge. These solutions were made with this purpose in mind. One wonderful thing that may happen if you apply this strategy is that you could make a model that works well with others. In this concept, artificial intelligence is responsible for activities including analyzing data, discovering patterns, and making predictions. Farmers, on the other hand, add their gut feelings, cultural knowledge, and capacity to change to the process (Castro, Iñacasha, Mesias, & Oñate, 2023).

The first step in the process is to gather data. This means putting sensors in the field to keep an eye on things like temperature, humidity, soil moisture, pests, and nutrient levels all the time. This is the first stage in the procedure. This is the initial step in the process, and it starts with this phase. This step is the first one in the process, and it is the first one that follows. All of these sensors work together to create a platform that is linked to each other. This platform lets you gather, sort, and deliver this information to central systems. Next, AI systems look at the data to find patterns, outliers, and connections that a person wouldn't be able to see in the same situation. This is done to make the identification procedure more accurate. Farmers may employ artificial intelligence to obtain help making choices that are not just timely but also well-informed. This help is given via the use of decision intelligence. The timing of irrigation, the choice of crops, and the control of disease are all examples of decisions that may fit into those groups. They are all instances of choices. Figure 1 shows a graph of the full process, from collecting data to carrying out the final solution. This is for your personal convenience. The goal of this research is to show how artificial intelligence systems may work with platforms, data streams, decision-making models, and actuation layers to make an agricultural environment smart and responsive. Accessibility, openness, and engagement are three important features of this system's human-centered design. These three traits are the most important ones that make up the system. It is really important that farmers can understand what artificial intelligence thinks, question their beliefs, and change them. Also, the interfaces should be easy to use, support several languages, and be able to work with people who have different levels of computer skills. Also, artificial intelligence systems are trained on data sets that are unique to the situation in which they will be used. This is done to make sure that the performance is good enough. Because of this, the systems can accurately adjust to a wide range of farming scenarios. Throughout the curriculum, this is done in a consistent way.

Figure 1. A human-centric AI framework in smart farming showing the flow from crop sensing to decision-making and actuation

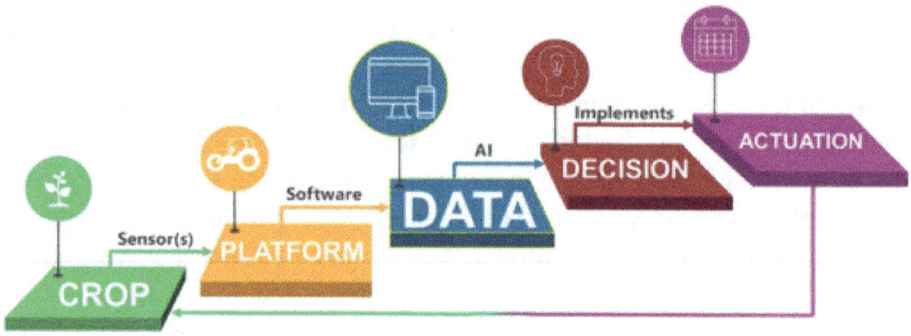

This method uses both human supervision and the process of actuation, which is the phrase for using automation to put judgments into action. The mechanism being spoken about is called "actuation." For example, AI could be able to send alerts or turn on automatic sprayers if it sees that there is a chance of a fungal epidemic. This would be an example of the ability in question. Also, it's probable that the farmer will still be the one to make the ultimate decision, based on a wide range of factors like checking the fields in real time, weather predictions, or fears about the economy as a whole. Along with this, AI can also help with managing resources in a way that is in line with sustainable norms. Farmers can get the most out of their inputs by looking at satellite photos, climate predictions, and historical yield data. This greatly reduces the damage they do to the environment and makes them far more resilient to the effects of climate shocks. Farmers may get all of these advantages. Because of this, they can reach all of their goals. Using smart farming platforms powered by AI is very important for cutting down on the amount of water utilized, the amount of fertilizer runoff created, and the amount of biodiversity protected. Here is a complete list of the main parts that make up the idea of regenerative agriculture. Using human-centered artificial intelligence has other advantages, such as making supply chain processes more efficient and open. Here are some of the benefits that artificial intelligence can provide. AI might aid a lot of fields, such market predictions and harvest logistics, and it could also make operations better and cut down on losses after harvest. Using artificial intelligence might help a lot of these sectors. Smallholder farmers, who frequently can't sell their goods on global markets, could be able to take advantage of the advantages that blockchain-integrated systems would provide. These advantages include more trust and the ability to collect revenue. Farmers can keep an eye on their crops from the time they are grown in the field to the time they are consumed by the consumer thanks to these technology. This helps them prevent any possible losses.

AI-powered chatbots and virtual advisers may provide assistance in real time and in the languages spoken in the various places that are affected by the crisis. This has to do with education and extension services for farmers. As a result, it is now feasible for specialized information to be available even in parts of the country that are not geographically close to each other. Artificial intelligence (AI) can fuel augmented reality (AR) technology that might provide people fully immersive training experiences. These systems may be used to mimic pest control, the development of irrigation methods, and the adaptability to climate. Because of the advent of digital technology, the agricultural industry has become more equal, and more people now have access to knowledge. But ethics will always be an important part of AI that is focused on people. People are the main thing we pay attention to. To deal with the problems that come with relying on technology, algorithmic discrimination, and protecting personal data, we need to pass laws that are open to everyone, use participatory design, and set up clear rules for how things work. Farmers must be allowed to use their data in a manner that is fair and equitable, and they must control their data and know how it is used. Artificial intelligence systems must show respect for the cultural norms, traditional knowledge, and gender roles that exist in agricultural communities. This is highly important. Only after that can these systems be called inclusive. One of the most important parts of Agriculture 5.0 is the effort to bridge the digital gap between generations. In order to make sure that the advantages of artificial intelligence are not limited to areas with more advanced technology, but are instead spread over the world, infrastructure has to be built. This will make sure that AI's benefits aren't confined to places with better technology. Providing cheap technology, teaching people how to use it, and bringing broadband internet to remote regions are all examples of costs that fall into this group. There are many various ways to speed up this change, such as using public-private partnerships, open-source platforms, and grassroots innovation hubs. The goal of this research is to look at the many ways, chances, and problems that come with adding human-centered AI to smart farming systems. The study will focus on a number of different facets of artificial intelligence. It looks at the following situations in a more detailed way:

- The evolution from Agriculture 4.0 to 5.0
- Core principles of human-centric AI design
- Use cases in precision farming, pest management, and agri-logistics
- The role of AI in sustainability and climate resilience
- Ethical frameworks for responsible AI deployment
- Strategies for inclusive implementation and community co-creation

In doing so, we position AI not as a replacement for the farmer, but as a digital companion that amplifies human insight, respects ecological boundaries, and contributes to a fairer, more resilient food system (Concepcion et al., 2022). Agriculture 5.0 is not only about smart technologies—it's about smart values: cooperation, empowerment, and stewardship. By placing humanity at the heart of innovation, we can ensure that the future of farming is not only intelligent, but also just, inclusive, and regenerative.

2. REVIEW OF LITERATURE

To offer an overview of the present position of human-centered artificial intelligence (AI), including the limits it encounters and the probable courses it may pursue in the future, the objective of this article is to provide an overview of the current situation. In the framework of smart farming, which is an essential component of the forthcoming Agriculture 5.0, which will take the place of Agriculture 4.0, this is carried out. In recent years, there has been a growing trend in the agricultural sector toward complete automation, which often leads to a decrease in the quantity of human involvement. (Fraser, 2022) This pattern comes to mind the fourth industrial revolution, which took place some centuries ago. When it comes to agricultural settings, the application of this method has been limited for a number of distinct reasons at this point in time. The inclusion of human knowledge and experience, which is often referred to as "human-in-the-loop," is usually found to be important. This is the case despite the fact that artificial intelligence is well acknowledged for its ability to execute tasks that are traditionally carried out by people. Moravec's Paradox, which says that some actions that are easy for humans to do are difficult for artificial intelligence to undertake, lends support to this argument by demonstrating further that this assumption is correct. Furthermore, the legislative framework of the European Union places an explicit emphasis on the significance of human control over this tool. This is done in light of the social, ethical, and legal considerations that are related with artificial intelligence. The purpose of this research is to get a knowledge of how the most recent developments in human-centered artificial intelligence might possibly be used in agricultural operations. The investigation focuses on the most recent breakthroughs in this field. These technological breakthroughs are being made with the intention of enhancing agricultural yields, decreasing the amount of labor and resources that are lost, and accelerating the velocity of the supply chain from the farm to the end consumer. The primary focus of this discussion is on the potential for artificial intelligence to assist human decision-making, which would, in turn, facilitate the development of an agricultural and food system that is robust, efficient, and environmentally friendly. This discussion is being held in light of the recent catastrophic weather disasters that have occurred all over the world. In conclusion, a framework for Agriculture 5.0 will be presented as the last part of the presentation. The objective of this framework is to guarantee that the requirements, capabilities, and surroundings of the necessary human stakeholders are in accordance with the most current technical capabilities. The implementation of Agriculture 5.0 in the future is dependant on this method, which puts a focus on artificial intelligence technologies that are broadly accessible and understood, and which significantly extend the number of activities that people participate in. This approach is essential to the success of Agriculture 5.0.

Experimenting with new ideas, gathering information, and altering current systems are all becoming more significant endeavors, and the metaverse is developing and becoming an increasingly vital site for doing so. As a consequence of this, a rising number of individuals within the academic community are interested in gaining knowledge about the ways in which new technology might be used in farming in a manner that is advantageous to the environment. Because of this, something transpired as a consequence. This is something that you should always keep in mind, but it is of much more significance now that a greater number of people are dealing with the metaverse. Some of the new technologies that are helping to convert significant ideas in the sectors of agriculture, nutrition, and sustainability include digital twins, artificial intelligence, and augmented reality. These technologies are contributing to the transformation of these essential concepts. In order to do this, the purpose of this research is to analyze the ways in which contemporary technology is generating changes in the meanings of things that were previously understood (Gardezi, Adereti, Stock, & Ogunyiola, 2022). An investigation that was conducted not too long ago shown that the use of immersive augmented reality technology has the potential to be of great assistance in changing the way in which folks learn about farming and get engaged in the business. Virtual farming platforms, in particular, may provide students with the opportunity to acquire knowledge on a variety of subjects, including but not limited to the weather, crop diseases, and irrigation systems systems. The purpose of these platforms is to create the sense that we are really experiencing real life. When children participate in activities like these, they may have a better grasp of how to manage a farm. As a result of this, students have the opportunity to gain more information about the subjects that were covered in the previous discussion. Both students and farmers have the option to take part in activities that take place in a safe atmosphere, which allows them to make decisions that are based on accurate information while they are participating in those activities. The existence of these virtual settings, which are intended to provide students with the opportunity to learn via practical experience, means that this prospect is now a reality. In addition, the study highlighted how augmented reality applications may be used to assist individuals in their professional life in making choices that are better informed by superimposing digital information on top of genuine agricultural settings. Due to the findings of the investigation, this was a completely plausible option. Cameras that are able to perform augmented reality enable all of this to be realized. This is made possible by cameras.

Agriculture 5.0 (Ag5.0) requires a considerable change in farming techniques in order to solve the issue of food insecurity on a global scale. This shift is necessary in order to address the problem. The use of modern technology, such as robots, artificial intelligence (AI), machine learning (ML), and big data analytics, is responsible for bringing about this transition. An in-depth investigation of the many ways in which modern technologies, such as artificial intelligence (AI), machine learning (ML), and other cutting-edge technologies, might be used to address current and future challenges regarding crop management is being carried out in this study. The purpose of this investigation is to facilitate the transition to Ag5.0. Precision agriculture has resulted in a considerable shift in crop management, which is indicative of a significant progress in general (Hameed, 2018). This transformation has been brought about by the use of precision agriculture. Agricultural activities that date back to the beginning of time are the starting point for the development of this technology, which finally leads to the sophisticated qualities of Ag5.0. The change that is now taking place is being accelerated by a number of emerging technologies, including blockchain, cloud computing, digital twins, 6G, collaborative robots, the Internet of Things (IoT), and quantum technologies. These technologies are playing a crucial role in the acceleration of the transition. The study also emphasizes the ways in which advancements in machine learning and current agricultural technology are enhancing our understanding of crop growth as well as our ability to assess and manage it. Our capacity to analyze and manage crop growth is also being improved. This is a topic that is discussed at length during the course of the research. In addition to that, this research investigates a large number of real-world case studies that illustrate the use of machine learning and deep learning in the context of crop monitoring. Intelligent sensors, robots that are driven by artificial intelligence, and strong communication networks are the driving forces behind the next generation of digitalization and decision-making in the agriculture industry. These technologies are expected to revolutionize business operations.

Additionally, there is a significant amount of curiosity with regard to the possible uses of digital twins in the agriculture sector. In addition to the items that were covered before, here is more information. This implies that you are able to model crops, animals, and agricultural equipment in real time. With the assistance of digital twins, you are able to develop realistic assets in real time, which means that you may use them. It has been possible to do this ever since the development of digital twins. A great opportunity has been made available to the agriculture sector as a direct consequence of this. Farmers have the ability to maximize the use of their resources, accurately predict the effects that climate change models would have on their crops, and keep a careful check on how their crops are doing as a direct consequence of these factors. With this in mind, it is possible that any and all of these things will take place. As part of the research that has been carried out, an investigation has been carried out to evaluate the many ways in which they might potentially contribute to the reduction of waste, the cultivation of more crops, and the preservation of the environment via their efforts. On the basis of what has been shown, it is impossible to reject the relevance of these two things. It has been shown that artificial intelligence and the metaverse are related to dietary systems that provide people nutritional advice that is both personalized to their specific needs and delivered in real time (Montavon, Samek, & Müller, 2018). The availability of this information is of great assistance with regard to aspects of nutrition. For the field as a whole, this adjustment is a big step forward in technological development. Through the use of games, these systems are intended to teach individuals about the ways in which their eating habits may have an impact on their health and to make it more fun for individuals to modify their behavior toward healthier eating habits. In order to achieve this aim, we want to encourage individuals to choose meals that are better for the environment. The clients will be encouraged to choose meals that are healthier for the environment and do less harm to it with the assistance of these strategies. It is the opinion of a lot of professionals that applications such as this one may perhaps aid persons in getting a great deal more knowledge regarding nutrition, especially younger people who have grown up with technology. In today's world, a significant percentage of young individuals are affected by this phenomenon. Because these technologies have the potential to aid in the monitoring of food, there is a lot of interest in blockchain technologies that are compatible with metaverse systems. This is because these technologies have the ability to help monitor food inventory. One of the reasons why a growing number of people are utilizing these electronic gadgets is because of this particular reason. The fact that people are paying attention to these technologies is due to a range of variables that are contributing to this phenomenon. Individuals have the power to make choices that are beneficial to their health when the food supply chain is available to public scrutiny for the first time. The decisions that have been taken are not only ethically acceptable, but they are also advantageous to the community of people who are concerned about the environment. When the facts are presented in a straightforward manner, individuals have the power to make moral judgments. In many cases, the demonstration of the transparency feature is performed via the use of interactive dashboards or interfaces that are able to be seen in augmented reality. The outcomes of a number of studies that were carried out at educational institutions such as colleges and universities indicate that this exposure helps to improve the trust that already exists between manufacturers and the potential customers of their products. The fact that people are able to eat food in a way that is more favorable to the environment is a direct result of this.

This research lays an emphasis on the revolutionary potential that these technologies have in terms of boosting agricultural output and resolving challenges that are linked with the security of food supply all over the world from a global perspective. This essay takes a look at the advantages and disadvantages of implementing Ag5.0, while at the same time highlighting the revolutionary possibilities that these systems will bring about. We concentrate a significant amount of attention on future trends and research requirements, such as transdisciplinary approaches, regional adaptability, and improvements in artificial intelligence and robotics. This is due to the fact that Agriculture 5.0 is a mirror of the direction that the agricultural industry is moving. Ag5.0 is a paradigm change toward precision crop management, which in turn promotes agricultural practices that are driven by data and sustainability (Otieno, 2023). Ag5.0 signifies a revolution in the paradigm. With the implementation of Ag5.0, the objective is to raise output levels while simultaneously reducing the impact that agricultural activities have on the surrounding environment. It is only through the use of competent crop management that it is possible to translate the knowledge that they provide into judgments that are very potentially lucrative. The use of data has become an important component of contemporary agricultural operations as a result of the fact that it provides assistance to farmers in the process of making significant choices. The exponential expansion of smart farming may be attributed, in large part, to the new developments that have occurred in data management techniques. When objective data is gathered by sensors with the purpose of enhancing performance in terms of both efficiency and sustainability, there are apparent benefits that may be realized. These advantages include the possible improvement of performance. Information that has the ability to minimize the amount of resources that are squandered and the amount of environmental harm that is caused is needed by these managed farms in order for them to be able to boost their output. The use of robotic technology and artificial intelligence approaches is the foundation for data-driven agriculture, which serves as the foundation for future sustainable agriculture. For agriculture to have a bright future, this foundation is very necessary. This may be accomplished via the use of robotic technologies. Based on an analysis of each important step connected with the process, this research investigates the present level of advanced farm management systems. The study begins with the collecting of data in agricultural fields and concludes with the application of variable rate. This article's objective is to equip producers with the information they need to provide them with the knowledge they need to make the best decisions possible in order to save money while also conserving the environment and improving the way food is produced in order to meet the impending population boom in a manner that is sustainable (Panarin & Khvorova, 2022).

The ideas that are associated with Industry 4.0 are connected to the activities that are carried out in the agricultural sector via the use of a new paradigm that is referred to as Agriculture 5.0. The scenario known as Agriculture 5.0 serves as the main point of our scoping study, which also draws attention to the crucial role that Explainable AI (XAI) has played in the growth of this sector. A comprehensive investigation of 84 articles that were published between the years 2018 and September 2023 was carried out by us in accordance with the criteria that were defined by the Preferred Reporting Items for Systematic Review and the Meta-examination Scoping Review. Our research has shown that XAI has the potential to be used in Agriculture 5.0, and we have a thorough grasp of how it effects intelligent farming (Rudd, Roberson, & Classen, 2017). This is evident from the outcomes of our investigation. For the goal of providing a conceptual framework for the inclusion of XAI, the objective of this study is to put a special emphasis on the implications that this will have on the transparency of models and the trust that users have in them. Nevertheless, the literature disregards XAI presentations rather often, despite the fact that they have applications that are not found elsewhere. The fulfillment of this requirement and the provision of a resource that can be used by academics, professionals working in the field, decision-makers, and educators who are active in the area of environmentally aware and technologically advanced smart agriculture is exactly what we want to do.

In spite of the fact that these new technologies have a great lot of potential, academics are worried about the digital divide that is developing all across the world at the present time. The degree to which certain regions of the globe have access to high-speed internet, immersion technology, and technological expertise continues to vary dramatically from one place to another. When it comes to digital platforms, this is particularly true. This does not change the fact that the difference is still rather substantial. The flip side of the coin, however, is that this disparity is still fairly considerable all things considered. According to the conclusions of the article, in order to effectively close this gap, it is necessary to use design solutions that are more inclusive. Some of the strategies that are used by the organization include the utilization of instructional materials that are sensitive to the cultures of different nations, platforms that are open-source, and interfaces that are capable of operating with more than one language. Moreover, professionals are researching the ways in which the extension of food connection to virtual situations effects the mental and social health of people. At this very moment, this investigation is being carried out at this very moment. The investigation of this new area of study is now being carried out by a number of people who possess a great deal of expertise on the subject concerned. It is possible that those who take part in immersive events would have a sensation of being even more disconnected from the food systems that they are seeing. This is because immersive events are designed to give attendees the sense that they are really there at the event. This is the reason why this is the case. It's possible that a person will have a more profound comprehension of a certain subject when they are entirely immersed in that subject. On the other hand, subject matter experts stress the need of preserving physical connections to the production of food and establishing hybrid models that integrate digital information with knowledge obtained in the real world. By returning to the topics that were covered before in the talk, we are able to grasp what is being said now. This study is being conducted with the intention of providing individuals with a fundamental understanding of the ways in which the metaverse may influence the manner in which people consume food and the viability of agricultural systems over the course of time. A comprehensive analysis of the research collection is presented in the following description. The fact of the issue is that it bring to light a great deal of problems and makes it clearly evident that a number of firms need to work together in order to create things in a way that is equitable and to build things that can be developed upon. That is the truth of the situation. There is no way that the introduction of new technology will be sufficient to really progress the situation. However, in addition to this, it will need human-centered innovation that is focused on fundamental ideas such as honesty, the purity of the environment, and education. As a result of the region's ongoing expansion and development, a growing number of people who are paying attention are coming to the realization that this is the actual situation.

3. UNDERSTANDING HUMAN-CENTRIC AI IN THE AGRICULTURAL CONTEXT

Combining artificial intelligence (AI) with agriculture creates a great chance to address world issues like poverty in rural regions, food insecurity, and climate change consequences. Conversely, the development and use of artificial intelligence technologies have to be seen from a human-centered standpoint if these technologies are to be successful and sustainable over long run. This implies that farmers and communities actively engaged in agriculture should take front stage in innovation focus. Human-centered artificial intelligence in agriculture covers a more broad spectrum of uses than just productivity and efficiency. It stresses sustainability even as it also emphasizes diversity, ethics, openness, and empowerment. This strategy specifically ensures that technology enhances the ecological and human aspects of farming, which is crucial in rural areas with diverse and sometimes sensitive character.

Definition and Core Principles of Human-Centric AI

Artificial intelligence (AI) labeled "human-centric" and "human-centered" systems are those specifically meant to raise human capacities and well-being while also harmonizing with societal values like justice, equality, and environmental stewardship. Regarding the agricultural industry, this means ensuring that the following artificial intelligence instruments are in place:

Address **real agricultural problems**
Support **human decision-making** rather than replace it
Are **accessible, trustworthy, and fair**
Promote **ecological sustainability** and **ethical standards**

Principles of Human-Centric AI

Human Agency and Oversight: Artificial intelligence should enable farmers to make decisions based on correct information. Technologies must keep being tools that help people, not instruments meant to control or dominate individuals. Regarding irrigation, for example, decision support systems are needed to provide recommendations instead of acting on their own will without user involvement.

Fairness and Inclusivity: While developing artificial intelligence solutions that would benefit all agricultural communities is crucial, smallholder farmers, women, young people, and disadvantaged groups within the agricultural industry should specifically be given top priority. Accuracies in the data must be eliminated, solutions suitable on a global level must be avoided, and technology must be customized to fit the needs of the local community and the languages spoken there.

Transparency and Explainability: Farmers and other agricultural stakeholders, particularly in cases where such models affect major operations like crop choice or pesticide usage, depend on a deeper knowledge of how artificial intelligence models function. Those that grasp artificial intelligence better are more confident and can verify or challenge the results.

Accountability and Governance: Governments, developers, and users all equally have to be held responsible for the consequences of artificial intelligence and its application. Relevant parts of this issue include governance of data, ways for mending damage, and ethical monitoring.

Environmental and Social Sustainability: Artificial intelligence should be used to support long-term sustainability goals including reduction of greenhouse gas emissions, enhancement of biodiversity, and promotion of resilient agroecological methods. This knowledge helps artificial intelligence to be in line with sustainable development and climate-wise friendly agriculture.

Aligning AI with Farmer Needs, Ethics, and Sustainability

Human-centric artificial intelligence must address the needs, values, and experience of agricultural communities if it is to provide significant help to the industry. The following are some major techniques to achieving alignment::

Co-Creation and Participatory Design: Artificial intelligence systems for farmers need to be created with them; they are insufficient on their alone. Participatory methods ensure that innovations are anchored in local problems, independent of whether such constraints are related to the availability of markets, the fertility of soil, or the frequency of bug outbreaks. These include the following elements:

- Farmer workshops and pilot studies
- Feedback loops for continual improvement
- Local customization of interfaces and outputs

Ethical and Cultural Sensitivity

Agricultural AI must respect:

- **Data privacy**: Farmers' data (e.g., yield, land usage) must be securely stored and ethically used.
- **Land and resource rights**: AI should not exacerbate disputes or inequities over land ownership.
- **Traditional knowledge systems**: Indigenous and local knowledge should be integrated into AI models where applicable.

Enhancing Economic Viability

AI should deliver tangible benefits such as:

- Increased yield and income
- Reduced input costs (e.g., fertilizers, water)
- Better market insights and price forecasting

This ensures that technology adoption leads to **long-term viability** and **resilience**, especially for resource-constrained farmers.

Promoting Environmental Sustainability

Examples of sustainability-focused AI in agriculture include:

- **Precision agriculture**: AI-guided planting, watering, and fertilizing based on soil and weather data.

- **Early warning systems**: Pest/disease detection to minimize chemical use.
- **Climate adaptation models**: Predicting crop performance under different climate scenarios.

These solutions help build climate-smart farms that are efficient and environmentally sound.

Capacity Building and Digital Literacy

Access to AI is meaningless without the skills to use it. Training farmers, extension workers, and cooperatives in digital literacy ensures:

- Better tool adoption and trust
- Effective interpretation of AI outputs
- Local innovation and entrepreneurship

Public-private partnerships, NGOs, and government agencies have a role to play in inclusive capacity building.

Challenges and Considerations

Despite its promise, human-centric AI in agriculture faces several challenges:

Digital Divide: Limited access to smartphones, internet, or electricity in rural areas.
Bias in Data: Algorithms trained on data from large farms may not generalize to smallholders.
Lack of Local Context: Imported AI solutions may ignore ecological and cultural variability.
Trust and Adoption: Mistrust due to poor past experiences with tech or lack of perceived benefits.

Mitigating these requires inclusive policymaking, localized R&D, and multistakeholder collaboration.

Human-centric artificial intelligence (AI), which values people, the environment, and ethics more highly than it does efficiency and profit alone, offers a fresh approach to agriculture. By means of human values integration into the evolution and use of artificial intelligence, we may design systems that are not only clever but also inclusive, equitable, and ecologically sustainable.

4. RESEARCH METHODOLOGY

This paper used a research design combining several research approaches to assess the effects of three different types of artificial intelligence (AI) systems—namely, Traditional AI, Semi-Human-Centric AI, and Human-Centric AI—on important agricultural outcomes such productivity, efficiency, farmer satisfaction, and sustainability. This was done to assess the results of various artificial intelligence systems. Using deliberate sampling, thirty farms in all were chosen; subsequently, the artificial intelligence system they used helped them to divide those thirty farms into three equal groups. Farmers' degree of satisfaction was scored on a Likert scale with 10 points; the mean crop production was tons per hectare. We subjected both of these factors to quantitative study. Some of the tools employed to gather data for real-time agricultural monitoring were sensors, drones, and satellite feeds. Formal questionnaires and interviews were used as another means of gaining qualitative information. Four main areas—performance outcomes (yield and satisfaction), task-based improvements (efficiency and error reduction in pest control, irrigation scheduling, and yield forecasting), sustainability contributions (water use efficiency, fertilizer optimization, and emission monitoring), and ethical considerations—trans openness, data privacy, and bias mitigating—focused the evaluation framework. Primary performance results were yield and satisfaction. The evaluation system built upon each one of these fields. A comprehensive study of user preferences for user interfaces was conducted across mobile applications, real-time dashboards, and voice interfaces in order to get a better knowledge of accessibility and acceptance trends.

Descriptive as well as inferential methods were used to help with data analysis. Descriptive statistics with a significance threshold of $p < 0.05$ were used to provide a whole picture of performance metrics across all three AI groups. Furthermore investigated were the presence of notable variations in the degrees of satisfaction and yield found using a one-way analysis of variance (ANOVA). Furthermore employed in order to record user experiences, ethical questions, and viewpoints on the user-friendliness of artificial intelligence was theme analysis used on qualitative replies. The small sample size, short observation time, and possibility of regional prejudices in the research, however, restrict it. The study is not without restrictions even if it provides a thorough knowledge of the many contribution artificial intelligence makes in agriculture. Future research should thus consider the possibility of using a wider geographic scope and doing longitudinal studies to get a more in-depth knowledge of the ways in which Human-Centric AI systems influence long-term agricultural productivity, sustainability, and farmer empowerment.

5. ANALYSIS AND INTERPRETATION

To facilitate the change toward Agriculture 5.0, which stresses the synergy between present technology and human intuition, artificial intelligence anchored on people is absolutely necessary. Review of previous implementations shows that artificial intelligence systems with features allowing adaptive learning and user interfaces readily understood are more likely to empower farmers than to replace them. Though they are collecting data in real time from satellites, drones, and sensors, human judgment, tastes, and local expertise steer these systems. Apart from ensuring farmers stay leading innovators in the area of agriculture, the interpretation stresses how artificial intelligence (AI) improves human capabilities to raise accuracy, production, and sustainability. Table 1 compares three artificial intelligence (AI) models: conventional, semi-human, and human-centric. The foundation of the evaluation is two main indicators of agricultural performance: the average degree of farmer satisfaction and the average output—measured in tons per hectare. The findings show a definite upward trend in both yield and satisfaction even if artificial intelligence systems are more focused on human needs. Conventional artificial intelligence (Group A) produced the lowest volume given an average productivity of 4.3 tons per acre and a satisfaction level of 5.1. On average production of 6.4 tons per acre and an 8.6. satisfaction rate, Group C, the human-centric artificial intelligence, excelled.

Table 1. Yield and Satisfaction Comparison Across AI Models

Group	N	Mean Yield (tons/ha)	Std. Deviation	Mean Satisfaction	Std. Deviation
Traditional AI (A)	10	4.3	0.52	5.1	1.1
Semi-Human-Centric AI (B)	10	5.2	0.61	6.7	0.9
Human-Centric AI (C)	10	6.4	0.49	8.6	0.7
Total	30	-	-	-	-

Whether or whether these variations would be regarded as statistically significant under the given conditions was investigated by use of a one-way analysis of variance (ANOVA). The results demonstrate that the groups vary statistically significantly; the scores of 0.00001 for the p-value, 25.3 for the F-value, and 47.53 for the inter-group sum of squares (SS) point this way. This suggests that it is rather improbable that random chance produced the noted variances. With human-centric artificial intelligence having the greatest positive benefits, these results provide additional evidence that the kind of artificial intelligence model has a major influence on the quality of experiences farmers have and the quantity of agricultural production they generate. Table 2 shows the summary statistics of the analysis of variance

(ANOVA) used to assess the relevance of the variances in yield and satisfaction noted amongst the three artificial intelligence models. The study used contained the following elements: the F-value, the p-value, the degrees of freedom (df), the mean square (MS), and the sum of squares (SS). The F-statistic comes out to be 25.3 as the variation between groups (SS = 47.53, df = 27, MS = 2, MS = 23.77) is much larger than the variance within groups (SS = 25.35, df = 0.94).

Table 2. ANOVA Summary – Yield and Satisfaction Across AI Models

Source of Variation	SS	df	MS	F	p-value
Between Groups	47.53	2	23.77	25.3	0.00001
Within Groups	25.35	27	0.94		
Total	72.88	29			

The extremely low p-value (0.00001) supports even further the hypothesis that the means of the three groups show a statistically significant variation. The statistical data shown here supports the hypothesis that the change toward artificial intelligence in farming more oriented on the engagement of people is directly connected to higher agricultural productivity as well as improved levels of satisfaction among farmers. More precisely, it emphasizes how important artificial intelligence architecture and design orientation are to the process of drastically raising agricultural output. Investigating the application of artificial intelligence in three key agricultural activities: yield forecasting, irrigation scheduling, and pest management Table 3 examines the usefulness of artificial intelligence based on the increase in efficiency, the reduction in mistakes, and the pleasure of farmers. With a satisfaction rating of eighty percent, a fifty- percent gain in efficiency, and a fifty-five percent drop in mistakes, the irrigation scheduling system has demonstrated the most important advantages among all of them. Major benefits of pest treatment also include a forty-five percent increase in efficiency and a sixty percent reduction in errors, both of which help to explain a seventy percent satisfaction level.

Table 3. Task-Based Impact of AI in Farming

Farming Task	Efficiency Increase (%)	Error Reduction (%)	Farmer Satisfaction (%)
Pest Control	45%	60%	70%
Irrigation Scheduling	50%	55%	80%
Yield Forecasting	35%	50%	75%

Though it shows a considerably less quantity of data, yield forecasting offers a considerable value. Its satisfaction score is 75%; its error rate dropped by 50%; and its efficiency increased by 35%. All of the above described improvements are proof that human-centered artificial intelligence may raise operational performance in many different fields relevant to particular tasks. These artificial intelligence models help farmers in making quick and well-informed decisions in high-stakes agricultural contexts, therefore lowering inefficiencies and building trust and usability all of which are vital. Human participation and user-friendly system interfaces are given great importance in order to reach this aim. Table 4 shows how artificial intelligence supports the sustainability of agricultural operations by means of three primary indicators: improving the use of fertilizer, raising the efficiency with which water is utilized, and monitoring emissions. With sixty percent of the entire effect on the efficiency of water use, artificial intelligence Farmers have a seventy percent acceptance rate and it provides a thirty-five percent improvement over traditional methods. Second place goes to fertilizer optimization, which shows traits of 65% acceptance rate, a 30% improvement, and a 55% contribution from artificial intelligence. By contrast, for pollution monitoring there is a fifty percent contribution, a twenty-25 percent improvement, and a sixty percent adoption rate.

Table 4. AI's Contribution to Sustainability in Agriculture

Sustainability Metric	AI Contribution (%)	Improvement Over Traditional (%)	Farmer Adoption (%)
Water Use Efficiency	60%	35%	70%
Fertilizer Optimization	55%	30%	65%
Emission Monitoring	50%	25%	60%

In addition to being deployed to increase productivity, artificial intelligence is also being applied to satisfy environmental objectives and reduce the ecological impact that agricultural operations leave behind. There is a significant amount of support among farmers for artificial intelligence solutions, particularly when these solutions are built with accessibility, interpretability, and immediate practical advantages in mind. This support is shown by the relatively high adoption rates that have been achieved across all programs. Artificial intelligence that is focused on people is very necessary in order to bring agricultural operations in line with the more comprehensive sustainability objectives that are envisioned in Agriculture 5.0. Combining user-centered interfaces with real-time environmental input is the means by which this objective is accomplished.

6. RESULT AND DISCUSSION

The results show that farms using AI systems focused on humans see notable changes in many other areas, including the degree of efficiency, the quantity of agricultural output, and the accuracy of choices. Artificial intelligence-driven platforms that provide personalized recommendations for irrigation, pest control, or harvesting times—for example—have proven the capacity to reduce input costs without compromising the environmental balance. Furthermore helpful is the fact that using cooperative design during the development of AI solutions results in higher farmer acceptance rates. The way this goal is reached is the inclusion of technologically advanced solutions into actual agricultural activities. This is a fundamental element of Agriculture 5.0, which demands a paradigm change in which the evolution of knowledge technology is directed toward satisfying the need of people rather than the other way around. The data that have been provided here offer additional proof that, given artificial intelligence targeted on people, agricultural ecosystems might develop into entities more intelligent, sensitive, and inclusive. Autonomous drones, artificial intelligence-based crop monitoring systems, and artificial intelligence-based decision support systems are the three basic technologies driving innovation in smart farming nowadays. Artificial intelligence is the foundation of all technological innovations right now. Figure 2 shows a comparison of three basic technologies; one may get this here. Adoption rate, degree of accuracy, and degree of human engagement define three most crucial factors taken into account when evaluating any technology. This study aims to provide light on the degree of technical resilience any technology has within the agricultural ecosystem, along with their pragmatic integration and degree of user dependence. Artificial intelligence driven decision assistance systems have a degree of accuracy higher than ninety percent when compared to the other two. Given this, it would suggest that they are rather consistent in helping with challenging agricultural concerns like crop choice, output projections, and resource management. Interestingly, they also need around 90% of human input, which is the most needed. This suggests that they have a cooperative dynamic wherein the system expands human knowledge without so displacing it. Here specifically, the "human-in-- the loop" strategy used by Agriculture 5.0 is shown.

Figure 2. Comparative Performance of Core AI Technologies in Smart Farming

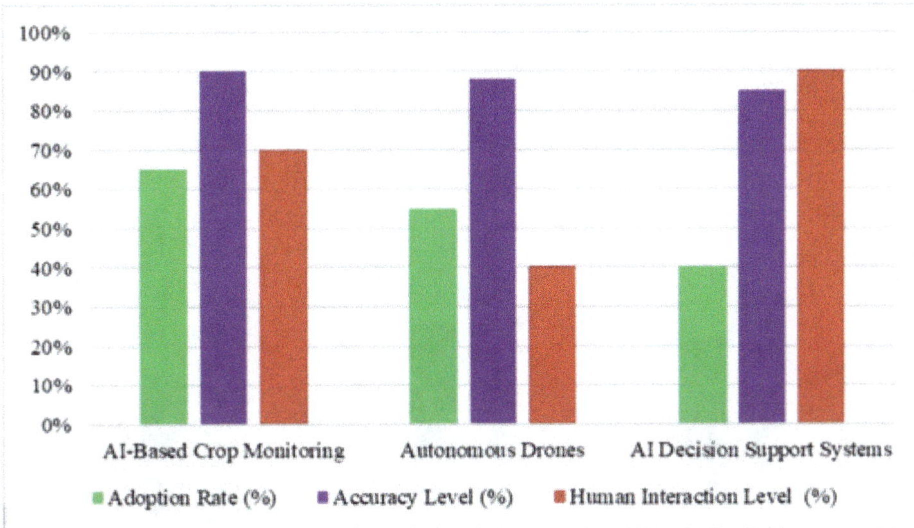

Though they are quite precise (almost 90%), autonomous drones have a percentage of acceptability significantly lower (about 55%) and a smaller degree of human involvement (almost 40%). This shows that, in spite of scientific data supporting them, they are not widely used in the everyday agricultural practices. Potential contributing factors to the issue might include end users' lack of knowledge with the technology, regulatory restrictions, and running expenditures. A crop monitoring system driven by artificial intelligence reaches a degree of accuracy of 90%, a level of human involvement of 70%, and an acceptance rate of 65%. This strikes a mix between the extremes. These data enable real-time information on crop health, insect detection, and soil conditions, thereby indicating a more balanced and easily available technology. Among farmers and companies, this technology is quickly being embraced. This gadget also offers help for the deemed reasonable human monitoring level. Three of the most crucial elements of artificial intelligence ethics in agriculture are openness, data privacy, and bias reduction; Figure 3 explores the dynamic interaction among ethical relevance, stakeholder understanding, and actual implementation in these three spheres. Having a one hundred percent score on the Importance Scale, safeguarding personal data privacy appears to be the most crucial objective. Given the fact that artificial intelligence systems are gradually gathering data unique to an area, exact, accurate, and frequently personal, this suggests that safeguarding farmer data is essentially of great significance. While farmers' awareness ranges around seventy percent, the actual implementation rate is rather sixty percent. Still, there is a great discrepancy in execution. This disparity suggests that there could be challenges such a dearth of technical infrastructure, regulatory enforcement, or farmer education.

Figure 3. Ethical Priorities and Practical Adoption in AI-Enabled Agriculture

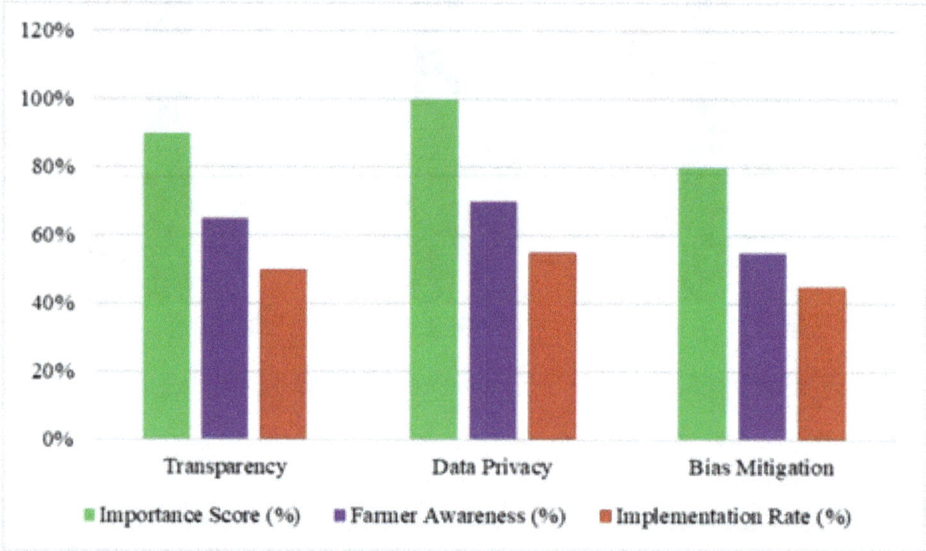

Transparency, the second-highest in importance at 90%, scores relatively well in Awareness (65%) but lags in Implementation (50%). This suggests that while stakeholders understand the necessity of explainable AI systems, there is still work to be done in making AI models more interpretable and accessible to non-technical users. Bias Mitigation, although recognized as important (80%), receives the least attention in terms of action, with only 55% Awareness and a mere 45% Implementation Rate. This shortfall raises concerns about the fairness of AI systems, especially in diverse agricultural settings where biased algorithms could inadvertently disadvantage smallholders or specific crop types. The data underscores the need for proactive strategies to address algorithmic bias before such systems become fully embedded in agricultural policy and practice. Figure 4 focuses on user-centric design in smart farming technology by comparing three interface solutions—Voice Interface, Real-Time Dashboard, and Mobile App Integration—based on Farmer Approval, Ease of Use, and Adoption Support. Mobile App Integration stands out as the most user-preferred format, scoring 90% in Ease of Use and 88% in Farmer Approval. Its high ratings reflect the growing smartphone penetration in rural areas and the convenience of having AI-driven insights readily accessible on handheld devices. Adoption Support also remains strong (around 83%), indicating that mobile platforms are well-supported by service providers and have infrastructural backing for widespread implementation.

Figure 4. User Interface Solutions in Smart Farming

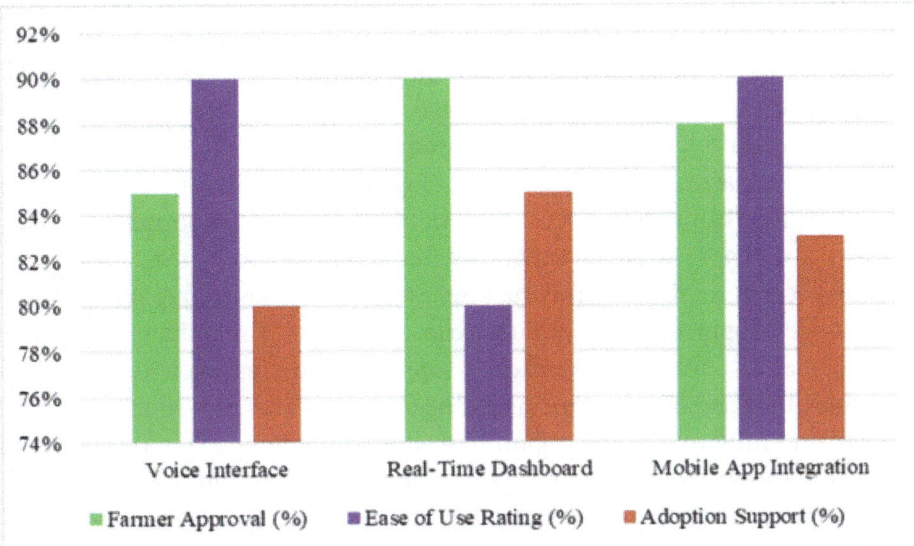

Real-time dashboards have a somewhat lower level of Ease of Use (80%) and a similar level of Adoption Support (85%), yet they are virtually as outstanding in terms of Farmer Approval (90%) and Adoption Support (85%). The lesser easy grade may result from either the difficulty viewing live data streams or exploring analytical tools. For farmers who hardly know about modern technologies, this is especially true. Still, dashboards are much valued as they allow for quick comments on farm activities. Voice Interactions not only appear to be promising in concept for their accessibility but also trail slightly in all three categories with 85% Farmer Approval, 90% Ease of Use, and 80% Adoption Support. This is especially true in settings with poor literacy and many language spoken there. These statistics indicate the necessity of further refinement in terms of the accuracy of voice recognition, language assistance, and real-world reaction even if they demonstrate a small degree of pleasure and support.

7. CONCLUSIONS

As the agricultural industry reaches the age of Agriculture 5.0, the application of artificial intelligence focused on people offers a huge possibility to reimagine agricultural systems not only technologically sophisticated but also ethically and socially conscientious. According to the findings of this study, artificial intelligence (AI) systems that give farmer engagement high priority and follow human values demonstrate far superior productivity, efficiency, and satisfaction than those others. Human-centric artificial intelligence empowers farmers to make educated choices, lowers resource waste, and encourages ecologically friendly practices, therefore transforming them from passive consumers of technology into important innovators. Still, many challenges need to be conquered like unethical conduct, data corruption, and limited access to digital infrastructure. Strong governance, regional artificial intelligence development, and continuous capacity building initiatives are key if one is to overcome these obstacles. Moreover, supporting co-creation with agricultural communities, transparency on artificial intelligence techniques, and ethical safeguards would assist to fully realize the opportunities of Agriculture 5.0 in the future. By raising its inclusivity, resilience, and future-oriented preparation, appropriate use of artificial intelligence targeted on people has the power to revolutionize agriculture.

REFERENCES

Belhajem, I., Ben Maissa, Y., & Tamtaoui, A. (2016). A Robust Low Cost Approach for Real Time Car Positioning in a Smart City Using Extended Kalman Filter and Evolutionary Machine Learning. *In Proceedings of the 2016 4th IEEE International Colloquium on Information Science and Technology (CiSt),* Tangier, Morocco, 24–26 October 2016; IEEE: Piscataway, NJ, USA, 2016; pp. 806–811.

Castro, S., Iñacasha, J., Mesias, G., & Oñate, W. (2023). Prototype Based on a LoRaWAN Network for Storing Multivariable Data, Oriented to Agriculture with Limited Resources. *In Proceedings of the Seventh International Congress on Information and Communication Technology, London,* UK, 21–24 February 2022; Springer Nature: Singapore, pp. 245–255.

Concepcion, R., Josh Ramirez, T., Alejandrino, J., Janairo, A. G., Jahara Baun, J., Francisco, K., Relano, R.-J., Enriquez, M. L., Grace Bautista, M., Vicerra, R. R., ... (2022). A Look at the Near Future: Industry 5.0 Boosts the Potential of Sustainable Space Agriculture. *In Proceedings of the 2022 IEEE 14th International Conference on Humanoid, Nanotechnology, Information Technology, Communication and Control, Environment, and Management (HNICEM),* Boracay Island, Philippines, 1–4 December 2022; IEEE: Piscataway, NJ, USA, pp. 1–6.

Fraser, A. (2022). 'You can't eat data'?: Moving beyond the misconfigured innovations of smart farming. *Journal of Rural Studies*, *91*, 200–207. DOI: 10.1016/j.jrurstud.2021.06.010

Gardezi, M., Adereti, D. T., Stock, R., & Ogunyiola, A. (2022). In pursuit of responsible innovation for precision agriculture technologies. *Journal of Responsible Innovation*, *9*(2), 224–247. DOI: 10.1080/23299460.2022.2071668

Hameed, I. A. (2018). A Coverage Planner for Multi-Robot Systems in Agriculture. *In Proceedings of the IEEE International Conference on Real-time Computing and Robotics (RCAR),* Kandima, Maldives, pp. 698–704.

Montavon, G., Samek, W., & Müller, K. R. (2018). Methods for interpreting and understanding deep neural networks. Digital Signal Processing: A Review Journal, 73, 1–15. doi: .DOI: 10.1016/j.dsp.2017.10.011

Otieno, M. (2023). An extensive survey of smart agriculture technologies: Current security posture. *World Journal of Advanced Research and Reviews, 18*(3), 1207–1231. DOI: 10.30574/wjarr.2023.18.3.1241

Panarin, R. N., & Khvorova, L. A. (2022). Software Development for Agricultural Tillage Robot Based on Technologies of Machine Intelligence. *In Proceedings of the International Conference on High-Performance Computing Systems and Technologies in Scientific Research, Automation of Control and Production,* Barnaul, Russia, 20–21 May 2022; Springer: Cham, Switzerland, pp. 354–367.

Rudd, J. D., Roberson, G. T., & Classen, J. J. (2017). Application of satellite, unmanned aircraft system, and ground-based sensor data for precision agriculture: A review. *In Proceedings of the 2017 ASABE Annual International Meeting;* American Society of Agricultural and Biological Engineers, Spokane, WA, USA.

Chapter 6
Enhancing Smart Agriculture Through Digital Twins:
A Comprehensive Analysis

Dipesh B. Pardeshi
Sanjivani College of Engineering, Kopargaon, India

Prachi Lokhande
https://orcid.org/0009-0007-2869-922X
Sanjivani College of Engineering, Kopargaon, India

Pranali Gaikwad
Sanjivani College of Engineering, Kopargaon, India

Sanika Pandit
Sanjivani College of Engineering, Kopargaon, India

Mudassir Khan
https://orcid.org/0000-0002-1117-7819
King Khalid University, Saudi Arabia

Shashi Bhushan
Universiti Teknologi Petronas, Malayisa

ABSTRACT

One of the biggest advances in smart agriculture is digital technology mixed with precision farming. Digital Twins (DTs), virtual representations of physical systems that enable real-time monitoring, modeling, and process optimization, have revolutionized agriculture. This article examines how distributed technologies (DTs), cloud computing, IoT devices, and AI-driven analytics might enhance agricultural productivity, resource efficiency, and farm management. Statistics utilizing analysis of variance (ANOVA) and post-hoc testing show that digital twin-based farming is superior than traditional and IoT-based farming in production consistency and environmental responsiveness. When sensor data resolution was increased by 5900%

DOI: 10.4018/979-8-3373-7077-4.ch006

ht © 2026, IGI Global Scientific Publishing. Copying or distributing in print or electronic forms without written permission of IGI Global Scientific Publishing is prohibited. Use of this chapter to generative artificial intelligence (AI) technologies is expressly prohibited. The publisher reserves all rights to license its use for generative AI training and machine learning model development.

and system variability was decreased, all key crop types demonstrated 20% to 36% output gains. Histograms, violin plots, and boxplots demonstrate DT systems' accuracy and stability. The study shows that digital twins enhance sustainable agriculture.

1. INTRODUCTION

There has always been a mix of science and instinct in farming. Farmers have used their knowledge, observation, and tradition to decide when to sow, how to water, and which crops to rotate for hundreds of years. But as we go into the 21st century and face more and more global problems, including climate change, soil degradation, rising populations, and unstable supply chains, the agriculture industry has to transform quicker and smarter than ever before (Barn, 2022). The digital twin is a new kind of electronic companion that is coming out during this critical change. A digital twin is not only a model on a computer. It is a virtual model of a real system, such a crop field, a greenhouse, a tractor, or even a whole farm. It can learn and respond to changes. It works by getting data all the time from sensors that are built into real-world assets. These data streams are processed and presented in a manner that lets people run simulations, guess what will happen, and make things work better. In farming, this may mean changing how much water you give your plants depending on how wet the soil is right now, using plant growth data to guess when the harvest will be, or using drone pictures and AI to find bug outbreaks (Cesco et al., 2021).

Figure 1. The general illustration of how the digital and physical worlds are dynamically integrated in a DT design

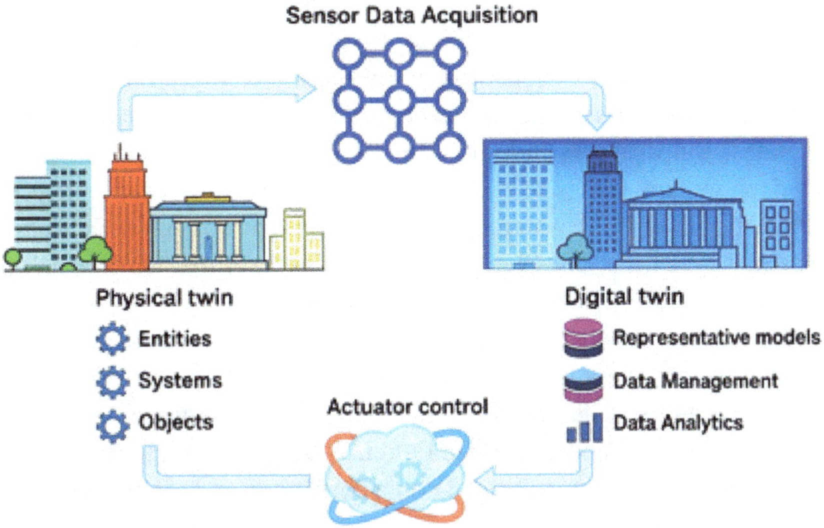

Figure 1 shows that the first step in the process is getting data. Sensors in the real world can pick up things like temperature, humidity, pH levels, fertilizer density, machine vibrations, and crop height. This sensor data goes into a digital twin, where data management tools store, sort, and get it ready for analysis. Analytics engines then turn this raw data into useful information, such as patterns, anomalies, trends, and suggestions for what to do next. You may use these insights not just to keep an eye on things but also to manage actuator systems. For example, you can switch on irrigation pumps, change the vents in a greenhouse, or send out autonomous equipment. The digital twin that comes out of this is an intelligent copy of the farm in motion. There are almost no limits to what it can do. Farmers may use it in precision agriculture to keep track of and manage changes in the field in real time, making changes per macrozone. Digital twins can mimic how animals eat, their health markers, and their breeding cycles in livestock management. They can predict delays in logistics and mimic inventory levels in supply systems (Edemetti et al., 2022). It's not only about being efficient; it's about being strong. When real-world systems are reflected in a controlled virtual environment, stakeholders may see what's coming and change their plans before things become worse. Digital twins do more than just copy things. They act like. And here is where their actual strength comes from: their capacity to simulate. Farmers may try out different techniques in the digital model before doing anything in real life. This lets them see what could happen without any risk. Should they use additional fertilizer or convert to a drip irrigation system? Will adding a new kind of crop change the equilibrium of the soil? If the temperature goes up 2 degrees, what happens to the yield? The digital twin can predict these scenarios and help you decide what to do next.

Making decisions is the most important part of smart farming. And excellent information is what makes decision-making work. Averages are widely used to make judgments about farming, such average rainfall, average output, and average fertilizer rates. But averages hide how things may change. Digital twins, on the other hand, show how things may change. They let farmers observe their work in great detail, such how much water a crop requires and how much each square meter of soil absorbs over time. They let you intervene in real time, which cuts down on guessing and makes things better. Digital twins are transforming the way people engage with farms, not only because of their technological features. They are being used to teach new farmers, educate customers, and develop confidence in supply networks. For example, a learner in Kenya may wear a virtual reality headset and look around a real-time model of a rice field in India. A person in Berlin may use a digital twin interface to scan a QR code on a coffee bag and see the whole journey of the beans from the Brazilian farm. This technology is changing how we think about openness, traceability, and learning (Fabbri et al., 2023). It also backs up a new way of working together. In traditional contexts, agricultural knowledge is generally limited to a small area and is based on experiences that are peculiar to that area. But with digital twins, this information may be shared. Digital twin models let two farms on separate continents exchange tactics and outcomes when they are both at danger from the same kind of weather. Extension personnel may stroll around a farmer's digital twin from a distance and provide them personalized advise without being there in person. These virtual connections may be the difference between being strong and falling apart when things go wrong as the world gets better at connecting.

Environmental stewardship is another area that is changing. Agriculture is a major reason for greenhouse gas emissions, water usage, and changes to land. Digital twins provide you a set of tools to lessen these effects by helping you prepare better and get feedback. A farmer may utilize their twin to try out alternative ways of rotating crops or planting cover crops and see how these changes effect the amount of carbon in the soil or the danger of erosion. You may make the most of water resources down to the level of individual plots. You can keep an eye on and balance nutrient cycles. This shows a change from reacting to the environment to taking action to protect it. People are also coming up with new ways to handle money. Farms may provide lenders, insurers, and investors real-time information on productivity, risk exposure, and sustainability initiatives by creating a digital twin. This makes it possible to provide more personalized loans, performance-based incentives, and insurance plans that can handle climate change. Farms that used to work in the dark may now be open and honest players in global food chains. But the change isn't without its problems. For one thing, the hardware and software ecosystems that make digital twins possible are still not fully connected. Many farms have old equipment or poor internet access, which makes it hard to broadcast data in real time. Data privacy and ownership are other problems. As farms become digital, they need to question, "Who owns the data?" Who gains from the information? To preserve farmers' rights and make sure that agricultural data is used in a fair way, policies and laws will need to change. Building a digital doppelganger is not just about technology; it's also about how you think. You need to look at the farm as a system of interconnected parts, each with its own behavior and way of reacting to change, not just a list of things to do. It implies being open to trying new things, always becoming better, and thinking about the future. This change has to be backed up by training, incentives, and community involvement for farmers who are already struggling with a lack of workers, climate stress, and market instability (Grieves, 2012).

Interdisciplinary cooperation is also important for the promise of digital twins. There is a job for agronomists, software engineers, economists, ecologists, and social scientists. Farmers and developers need to work together to construct digital twin projects, and they need to combine native knowledge with scientific data. They also need to be open to everyone. To sum up, digital twins are not simply things that exist in technology; they also exist in society and the environment. Digital twins might become a big part of how we all deal with the problems that food systems will face in the next decades, such as harsh weather, population increase, and changing diets. They let you see issues before they happen, try out remedies before you use them, and keep track of progress in real time. They make the unseen things that affect our farms clear and give voice to the quiet forces that shape them. We need to make sure that all kinds of farmers can use digital twins, not just those with big farms or tech-savvy teams, so that they can reach their full potential. Platforms that are open source, sensors that are cheap, and training models that are based on the community will be very important. We also need to agree on what sustainable and fair farming looks like in the digital era and how digital twins can help us get there. This in-depth study will look into digital twins in smart agriculture from several angles, including the technical, human, economic, and environmental ones. It will look at case studies, essential enabling technologies, policy implications, and a plan on how to make sure everyone can use it. In this way, it hopes to provide a thorough and detailed look at how digital twins might improve agricultural systems of all sizes, from local plots to worldwide networks.

2. REVIEW OF LITERATURE

Digital twins are duplicates of real-world items, systems, and processes. We call them "digital twins." They duplicate these features by using their ability to imitate and guess how others will act in the actual world. These technologies make it possible to develop virtual farm simulations that take into consideration factors like the weather right now, the kind of soil, and the amount of crops that will grow. This is helpful for smart farming since it enables you do these tests (Jo, Park, Park, & Kim, 2018). Farmers may now obtain a real-time, changing view of their fields owing to soil, plant health, and environmental sensor networks, weather forecasts, and high-resolution satellite and unmanned aerial vehicle (UAV) photos. Because of this, they can quickly and properly decide on crucial things like when to water, how to deal with pests, and how to fertilize. All of these things assist to make farming more productive and long-lasting. This article is about the study of digital twins for smart farming. This study will look at a lot of things, such as crop modeling, precision farming, and the tools that go along with it. This article talks about the study on digital twins, what they may be used for, and what such usage could signify for farming. It also speaks about things like how to integrate data, how to keep data private, and how vital it is to have good data in order to make correct forecasts and simulations. At the conclusion of the presentation, there is a look at the future of agricultural research and professional development, as well as a talk about the many ways that digital twins may be employed.

Because of its roots, farming has always relied heavily on gut feelings, experience, and a strong connection to the land. But farming is going through a tremendous upheaval right now because of climate change, population increase, and problems with the environment. Digital twins are a new kind of technology that combines old-fashioned methods with the most recent scientific research. This is the main reason for this adjustment. Farmers can now keep an eye on, simulate, and make better choices with more assurance than ever before, thanks to these cutting-edge technologies that are practically real-time virtual representations of actual agricultural landscapes. Think about being able to view a complete digital reflection of your farm on a screen. This mirror would show exactly what is happening in the soil, crops, water systems, and even with the cows. It would get updates in real time via sensors, satellite feeds, and drone sightings. Digital twins are giving the farming business a whole new level of power. Digital twins aren't just static models or one-time simulations; they're live things that evolve as their surroundings do. Farmers can now keep an eye on the health of their plants from seedlings to harvests, check the moisture level of the soil in certain areas, and understand how the performance of a tractor affects the amount of food that is produced. This level of detail used to be impossible to understand, but it is now necessary.

Smart agriculture is the use of digital technologies like sensors, information technology, autonomous vehicles, data analytics, predictive modeling, and others in farming. It has the potential to improve food security, water use, fertilizer and pesticide use, and farm efficiency. Still, intelligent agricultural technology isn't that popular yet, and how much people like it depends on where they live and what kind of technology it is. The goals of this paper are to: (1) create a conceptual framework for digital twins and smart agriculture that takes into account the needs and features of farms; (2) show how the proposed conceptual framework can be used in a case study; and (3) talk about the problems and chances that smart agriculture is facing right now. We will start by setting up a framework that will be the foundation for making agricultural information systems (Jain et al., 2024). This paradigm is based on the infological method and can be broken down into four primary phases: collecting data, processing data, analyzing and evaluating data, and finally consuming information. After that, we'll utilize the framework to show and speak about how digital twins and smart agriculture may be used in the real world to help with nitrogen (N) fertilization for crops. This will happen in the field. Both the case study and the literature that was quoted stress how important it is to find the right amount of nitrogen fertilizer to use. Other factors that are taken into account are the land's geographical variability, the soil's properties, and the crop yield, as well as how these factors change over time. Finally, we look at the possibilities and challenges that lie ahead, focusing on areas where the average farm size was rather small to begin with. We think that the huge amounts of data that digital twins may gather might be utilized for research that is both stable and predictive. If this research is used, it might help farmers and the agricultural sector maintain its social, economic, and environmental sustainability.

This technology is so powerful because it can be used in so many different ways. For instance, in precision agriculture, digital twins help farmers apply water and fertilizer more accurately. They allow for site-specific treatments instead of treating whole fields the same way. This not only saves money, but it also protects the environment and boosts output. When greenhouses have digital copies of the weather, farmers may change the levels of CO2 and humidity in the air. You may check changes electronically before they are made in real life. This cuts down on waste and makes sure that the growing conditions are always perfect (William et al., 2025). People who take care of animals, which some people think is hard to digitize, are also benefiting. Wearable sensors on animals provide information about their health, such as their temperature, heart rate, and activity patterns, to a central modeling system. If an animal shows signs of illness early on, the digital twin will quickly let the owner know. This will mean less discomfort, fewer breakouts, and surgeries that work better. Also, it opens up new ways to keep an eye on animal welfare, which is becoming a bigger issue for both consumers and managers.

These digital ecosystems also go beyond the farm and into the infrastructure of the agricultural supply chain. By giving farmers a better understanding of how global events affect their businesses, it is feasible to imitate logistics, market shifts, and delivery dates to help them make better business choices. One of the best things about digital twins may be that they make you see things in a new way. Because the data is presented in a way that is both clear and interesting, farmers start to see their work in a new light. You need to work harder, but you also need to work harder mentally. These tools make it easier for farmers to make tough decisions, so they can focus on strategy and stewardship instead of always having to deal with issues. They also encourage people to try new things. When you have a virtual model, you may try out different ways to plant or deal with pests without putting the real crop at risk. But the journey is not without its problems. One of the hardest things to do is to combine the huge amounts of data that are needed to make correct twins. Farm equipment, weather stations, soil sensors, and drones all talk to one other in their own digital languages. Integrating all of this data into a single model requires a complex infrastructure and a lot of technological know-how. These criteria could seem too much for smallholders, especially in underdeveloped nations. The fees are still rather high, and there isn't always enough training. This is a really important period when new ideas might speed up the process of inclusion, and partnerships between the public and commercial sectors will be extremely important to make sure that everyone has equal access.

Digital twins might make smart farming more productive and sustainable in a big way. A "digital twin" is a virtual object that acts like a real-world item. Digital twins are very important for managing farms because they let you separate planning and controlling physical flow from the actual motions of the farm (Yogeesh, Girija, Rashmi, & William, 2024; Sharma et al., 2024; William et al., 2023a). Farmers may now keep an eye on what they're doing from far away by utilizing digital data that is practically real-time. This means that there is less need for people to work and be supervised on site. There has been a big improvement over what was there before. Because of this, they can quickly respond to changes that have been planned for and judge actions based on what has really happened. This article looks at how digital twins may be used in smart farming and talks about the subject in depth. The word "digital twin" is defined in this article, and a list of distinct sorts of digital twins is given. It gives a theoretical basis that might be useful for making and using digital twins. The framework uses an idea that is common to all system control. The framework also includes a reference design and implementation method for Digital Twin systems that is based on the Internet of Things at the A level. The European Internet of Things 2020 initiative is working on five main smart agriculture application cases. These use cases include running an agribusiness, a dairy farm, a greenhouse for growing plants, an organic vegetable farm, and taking care of animals. Use these examples to put the framework to the test and see how well it works (William et al., 2023b; Monteiro et al., 2018; Skobelev et al., 2021; Zermas, Morellas, Mulla, & Papanikolopoulos, 2018).

There also needs to be better data standards. As more people enter the sector, it is becoming harder to exchange ideas and make good solutions work on a larger scale since there are no common standards. Some of these stakeholders include colleges, IT companies, and farms. The promise is not completely kept if the digital twin of one greenhouse can't talk to the irrigation system of another greenhouse. The current body of research is not well organized, even if it is full of new ideas. Most of the study looks at extremely successful pilot projects or smaller applications, but very few look at how well these systems will work in the long term or how well they can grow to a global level. There is still a lot of hope for the future, however. When you add AI to digital twins, the system will be able to make far better predictions. Think of a system that not only keeps track of a crop but also learns from how it grows each year and uses this knowledge to provide ideas for how to make things better and to warn you about any problems before they happen. Edge computing will make it feasible for a lot of this analysis to happen on-site, which will speed things up and allow for real-time responses even from far away. Some developers are even trying out blockchain technology to make the data of digital twins entirely transparent and impossible to change. This might revolutionize the way we track food supply chains for good. During this change, a lot of focus will also be put on education and training. Digital twins make it easier to keep an eye on crops and animals. They might also be used to help students and new farmers learn in a hands-on way. Users may learn about best practices, how to deal with frequent problems, and eco-friendly alternatives to chemical-heavy treatment procedures by using simulated farming situations. These models are also easy to get to from anyplace, which makes it simpler for individuals to exchange information across national and community boundaries. This helps fill up the gaps that most conventional agricultural extension initiatives leave.

Digital twin (DT) technology might help various fields, including farming, be more productive, make better choices, and be better for the environment. This research looks at how DT technology may be used in farming, what difficulties it might cause, and what other applications it can have. This research focuses on layering and categorization to look at the core notions and structure of DTs. This study looks at how decision trees (DTs) may help with farming decisions, such when to sow crops, how to care for animals, how to deal with pests and diseases, and how to utilize resources and tools to produce the greatest outcomes. We also talk about how tactics for gathering, evaluating, and modeling agricultural data may aid with the use of information technology. Some of the technological issues that make it hard for DTs to be extensively used in agriculture include gathering data, combining data, and making three-dimensional crop models that are all the same. It has a lot of potential, however there are some technical issues with it. We'll end by talking about what DT technology could be able to achieve in the future and why it's necessary to get away of the constraints that are now in place so that it can keep operating and be utilized more broadly.

Digital twins seek to improve operations in many areas of the economy by linking the digital and real worlds. This research looks at how digital twins may be used in forestry, animal husbandry, and open-field agriculture. This study shows how digital twins are used in many different fields. It is important to think about the following: how effectively the digital twin is integrated and how mature it is, how data is collected, how technically skilled the people are, and what the input and output are like (Sharma et al., 2024). This article looks at digital twins, how they may be used together, and the biggest problems and opportunities in the fields that are being studied via the use of four main research areas. Scientists have found that a complete digital twin is still a long way off, even though a lot of development has been achieved. They got to this decision after looking at the matter. Even if there are still problems to solve, it seems that the most important thing is to include the professional talents of a lot of people who are involved. There is also a new definition of digital twins that is right for a certain company. This definition was made because of the problems that were found when the present study was looked at. When this theory is put into action, it would take into consideration the unique features of complex biotic and abiotic systems. It is reasonable to expect that the research's findings will be useful to anybody who is interested in the topic. It is expected that this research will lead to more people being aware of the huge benefits that digital twins provide. It is also expected that this study will encourage a more thorough and in-depth look at this new field.

Digital twins are not just a new technology; they are a completely new way of thinking about farming. The industry has changed from being reactive to proactive, from being based on guesswork to being based on understanding, and from being broken up to being whole. They provide farmers a place to share their expertise and experience, as well as statistics that they can use to improve their work. They provide researchers and engineers a place to work where they may test their ideas on a large scale. And for society as a whole, they represent a significant step toward food systems that are more sustainable, as well as more efficient and open. But we need to be careful as technology grows. We need to remember the people who ran the fields as we make these digital copies. If a system is too difficult, too expensive, or too exclusive, it won't perform what it was meant to achieve. To be successful, you need to work with farmers, not just for them. We need tools that can work in a lot of different languages, operate on cheap devices, and be useful on farms of all sizes. Part of the equation is protecting privacy, making sure that data is used in a fair way, and finding a balance between innovation and equality. For now, the literature has given us a comforting picture of what is possible. Even said, the field is still new, and there is a lot that hasn't been looked into yet. Digital twins might be the basis for Agriculture 5.0 and beyond. They can do anything from predicting crop yields in real time to making decisions automatically. But they will only be able to take us where we need to go if they are easy to get to, adaptable, and built with the needs of farmers in mind. The digital twin is more than just a tool in this new era of smart farming; it is a dependable partner that is showing the way forward.

3. SMARTER FARMING WITH IoT AND DIGITAL TWINS

The use of digital technology in the agricultural industry is fundamentally changing the accepted agricultural methods applied for millennia. Digital technology makes it possible to make decisions grounded on data, operational efficiency, and sustainable resource management as well as in other aspects. Among the most important developments in this field is the Digital Twin, a virtual simulation of real-time updated actual assets, systems, and processes. Digital twins can do tasks in the agriculture sector that were unthinkable only years ago. These exercises include precise resource management, irrigation optimization, environmental result prediction, and crop condition simulation. The most significant feature of this technology is its capacity to replicate, in real time, within a virtual reality, the circumstances and changes that take place on the real farm. Through constant data collection, transmission, and analysis, cloud-based platforms, edge computing, and sensors linked to the Internet of Things (IoT) combined provide these capabilities.

Figure 2 shows clearly that a network of physical objects placed on the farm forms the foundation of the digital twin ecosystem in smart agriculture. Among the elements that define these features are edge computer units, satellite pictures, weather sensors, drones, and soil sensors. These components gather some of the agronomic elements like temperature, humidity, and soil moisture. Furthermore in charge of compiling essential information on crop health are these components. Following that, the data is sent to cloud platforms where it is examined, combined, and used to create a dynamic model of a virtual farm arrangement. The graphic shows how seamless visualization, simulation, and predictive modeling made possible by real-time data from the field to the cloud. These qualities help farmers to make important choices.

Figure 2. Digital Twin Integration in Smart Agriculture

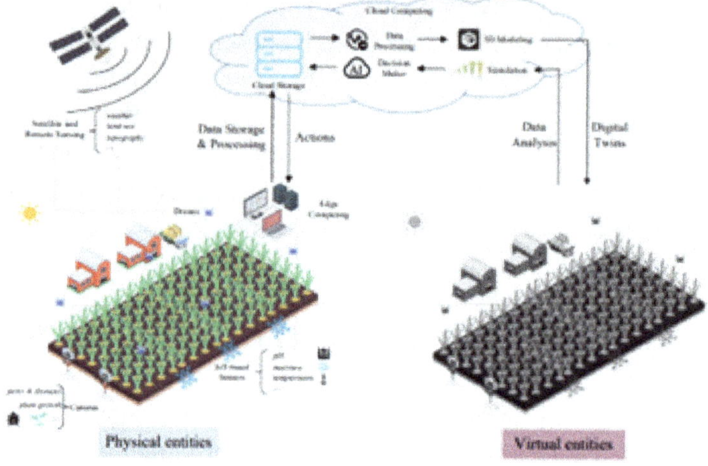

The first step in combining IoT with digital twins is to gather data from the real world. Sensors that measure temperature, humidity, pH levels, nutrient concentrations, and other important factors are built into the fields. Drones and satellites can see things from above, such changes in crop health, insect infestations, and irrigation coverage. Edge computing nodes provide this raw data to centralized cloud systems, which lets users get changes in real time with very little delay. Once data gets to the digital infrastructure, it goes into the data processing layer. There, machine learning engines, artificial intelligence, and complex algorithms start to make sense of it. This step involves cleaning, storing, and organizing data before turning it into useable information. Digital twins are accurate virtual copies of farms, equipment, animals, or even individual plants that show these insights in a way that makes sense.

These digital models are not just pictures of things. They can simulate future situations, guess what will happen, and suggest particular actions to take. For instance, a digital twin may predict how a drought would affect a certain area, show how different fertilizer methods will affect yields, or tell you when the best time to use pesticides is. This feature lets farmers test their choices in a virtual setting before putting them into action in real life, which lowers risk and boosts efficiency. The application layer is another important layer that connects farmers and farm managers. This is where data may be used. People may connect with their digital farms using smartphone dashboards, AR eyewear, or VR platforms. They can see which portions of a farm need water, keep an eye on the health of their animals from afar, or see what would happen if they planted a different crop type. The outcome is not just wiser farming, but also more intuitive, open, and well-informed choices. Figure 3 shows the tiered structure of this digital twin system, which gives a clear picture of how raw sensor data moves through the perception, network, and application levels to help smart farming decisions.

Figure 3. Layered Architecture of a Digital Twin System

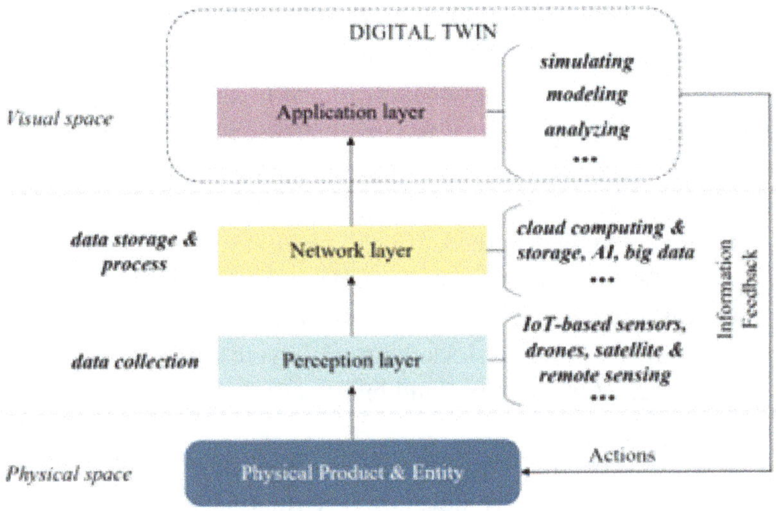

Using IoT and digital twins in farming has a lot of benefits. First and foremost, it makes precision farming better by enabling farmers use water, fertilizers, and pesticides precisely when and where they are needed. This not only cuts down on waste, but it also increases crop yields and protects the environment. Second, it offers predictive analytics, which helps farmers plan their harvests, figure out how much of a product is available on the market, and handle disease outbreaks. Digital twins also let you see and control farming activities in real time. For example, if the soil moisture level goes below a certain threshold, real-time sensor data might immediately start irrigation. You may also modify the settings of a greenhouse on the fly to maintain the optimal growth conditions, which helps crops grow better and consumes less electricity.

It's also crucial that these technologies be favorable for the environment. Digital twins fit with the goals of climate-smart agriculture because they help farmers use resources more wisely, reduce their carbon footprints, and promote acts that are good for the environment. They also make it easy to understand how food moves through the system and keep track of it. Because dual systems keep such good digital records, both customers and authorities can access complete histories of how food was grown, picked, and transported. But it's not always straightforward to set these things up. Infrastructure concerns, including not having internet access or power in rural areas, might impede down real-time data transfer. Smallholder farmers are still having trouble with the high initial costs of IoT devices and digital twin systems. Also, to protect farmers' rights and make them more comfortable with digital platforms, data privacy and ownership problems need to be addressed.

We will need a combination of government support, partnerships between the public and private sectors, and open-source tech solutions to move beyond these difficulties. Governments and non-profits may assist pay for training programs and developing infrastructure. Companies that engage in agri-tech may create affordable, modular solutions that are great for small farms. Schools could work together to make sure that all of their data security and standards are the same. The link between IoT and digital twins is expected to become stronger as 5G, edge AI, and blockchain become more widespread. 5G will make it simpler to connect in distant areas, which will make data transfer faster and more reliable. Edge AI will make it possible to process more data locally, which will lower latency and the requirement for centralized systems. Adding blockchain will make it possible to keep secure, unchangeable records of transactions and interventions, which will aid with digital certification and traceability.

Digital twins are becoming quite important tools for learning and teaching. Schools that teach agriculture are developing virtual farms for students to use to practice making decisions, look at real-time data, and estimate what will happen. These hands-on activities prepare young farmers and agricultural professionals for the digital revolution in farming. Digital twins are also assisting with improved inventory control, less waste of food, and better logistics in supply chain management. By replicating transportation routes, storage conditions, and market demands, they make food systems more flexible and able to respond to changes. Right now, IoT and digital twins are making farming smarter. It's becoming increasingly real, and it might revolutionize how we grow, manage, and share food. There are still difficulties, but steps are being taken to make farming smarter, stronger, and more sustainable in the future. You need to buy technology and vow to be open, fair, and share ideas if you want to fully accept this transformation. As we go toward Agriculture 5.0, digital twins and the Internet of Things (IoT) will be vital for feeding the globe and conserving the land.

4. RESEARCH METHODOLOGY

The purpose of this study is to explore the influence that digital twin technology has on smart agriculture by comparing its performance to that of conventional farming techniques and farming methods that are based on the internet of things. In order to ensure that the information obtained is comprehensive and trustworthy, an approach that utilizes a combination of methodologies was used. In this method, quantitative data analysis and statistical validation approaches were combined in a variety of ways.

Research Design

A comparative experimental technique was used in order to investigate the differences that exist between three distinct agricultural systems: traditional agriculture, agriculture based on the internet of things, and agriculture based on digital twins. All of the different approaches that were used have the evaluation of environmental adaptability, crop production performance, and sensor efficiency as their primary areas of concentration.

Data Collection

In order to obtain all of the necessary data, we used both modeling platforms and field research technologies. The data that was gathered in real time by sensors that were connected to digital twin systems and the internet of things (IoT) included the amount of moisture that was present in the soil, as well as the temperature, humidity, and light levels. For the purpose of ensuring that the findings are applicable to a wide variety of situations, yield data was collected from a wide variety of crop kinds and environmental circumstances.

Instrumentation

The installation of sophisticated sensor arrays and weather stations allowed for the collection of data with a high resolution and pertaining to agriculture and the environment. The capability of digital twin systems to imitate the dynamics of real farming via the use of simulation software and machine learning algorithms makes it feasible for the development of predictive analytics to take place.

Sample and Sampling Technique

A stratified random selection process was used by the researchers in order to choose a total of ninety agricultural plots, which were dispersed in an equal manner throughout the three different systems (thirty for each approach). During the course of the study, this was done in order to guarantee that a wide range of soil types, crop kinds, and climate zones would all be covered simultaneously..

Data Analysis Tools

The statistical analysis methods that were used included both descriptive and inferential approaches. We used analysis of variance (ANOVA) and post hoc pairwise comparisons in order to determine whether or not there were any significant variations in yield. Because of this, we were able to differentiate between the different categories and determine which ones were distinguishable from the others. Python and R are the two programming languages that were used in the development of visual analytics. Analytical tools such as histograms, bar charts, boxplots, and violin plots were included in this category. One of the goals of these visual analytics was to conduct an analysis of the patterns and variability of the data.

Validation and Reliability

It was able to establish its internal validity due to the fact that the study was conducted under controlled farming circumstances and used comparable measuring procedures throughout the whole process. As a consequence of conducting our research in a wide variety of field settings, we were able to demonstrate the external validity of our findings. Over the course of the growing seasons, a number of research were conducted in order to enhance the reliability and consistency of the findings.

Ethical Considerations

The research was carried out in accordance with ethical norms to the extent that it minimized the influence on the environment throughout the testing procedure, ensured the confidentiality of the data, and obtained informed permission from the farmers who participated in the study.

The study was conducted using an approach that was both rigorous and statistically sound, which made it feasible to conduct an accurate assessment of the value of digital twin technology in the farm manufacturing business. The findings of this research provide a solid basis for getting a better understanding of the revolutionary potential of digital twins in terms of increasing agricultural output, sustainability, and efficiency. Through the use of sophisticated analytics in conjunction with the acquisition of data from the real world, this objective may be accomplished.

5. ANALYSIS AND INTERPRETATION

There is a possibility that digital twins may be used in smart agriculture in order to make continuous data collection and simulation more feasible. The construction of virtual models that are strikingly similar to real agricultural scenarios is the means by which this objective is accomplished. By merging data from weather stations, crop monitoring systems, and sensors connected to the Internet of Things, digital twins are able to do in-depth research on the growth of plants, the conditions of the soil, and the effects on the environment. By analyzing both historical and present data, agricultural decisions may be made more rapidly and with a greater level of knowledge. This allows for the identification of patterns, the prediction of outcomes, and the use of most available resources. As a consequence of the implementation of digital twin technologies, particularly with the advent of smart agriculture, there has been a considerable shift in the manner in which data is collected, processed, and used in agricultural settings. It is clear from looking at Table 1 that the use of digital twins has led to an exponential rise in the number of sensor readings that are collected on a daily basis. Prior to the implementation of digital twins, soil moisture sensors collected an average of twelve readings per day. However, with the addition of digital twins, the average number of measurements collected per day climbed to 288. This is a two thousand three hundred percent increase. Furthermore, the frequency of data collection for the temperature and humidity sensors rose by 5900%, going from 24 measurements per day to 1440 readings per day on average. This is a significant increase.

Table 1. Sensor Efficiency Improvement in Smart Agriculture Systems

Sensor Type	Avg. Readings/Day (Before)	Avg. Readings/Day (After)	% Increase
Soil Moisture	12	288	2300%
Temperature	24	1440	5900%
Humidity	24	1440	5900%
Light Intensity	8	720	8800%

The most noteworthy improvement took place in the monitoring of light intensity, which saw an incredible 8800% increase in the number of daily measurements, going from 8 to 720. It was clear that this was the most significant area of progress. These modifications are necessary because they make it possible to continuously monitor the environment and gather data with a high resolution. The forecasting of diseases and pests, the optimal arrangement of irrigation, and the thorough assessment of the general health of crops are all facilitated by this. These digital twins are largely responsible for integrating and analyzing this data, which enables proactive decision-making rather than reactive decision-making. Real-time reaction is an essential component of smart farming, and digital twins are the primary players in this process. An analysis of variance (ANOVA) test was carried out (see Table 2) in order to verify that agricultural systems that are based on digital twins are more effective than those that are only based on the Internet of Things. An F-statistic value of 214.28 and a P-value of 2.43E-34, both of which are considerably below the significance threshold of 0.05, indicate that there is a statistically significant difference between the three farming systems. The data reveal that this difference is statistically significant. The sum of squares across groups (23.78) shows that the farming method that was applied was nearly completely responsible for the variance in agricultural output, as opposed to differences that were the result of chance. This is in contrast to the residual error, which was 4.83.

Table 2. Statistical Significance of Digital Twin-Based Farming (ANOVA Results)

Source of Variation	Sum of Squares	Degrees of Freedom	F-Statistic	P-Value
C(Farming_Method)	23.77723445	2	214.2778988	2.43E-34
Residual	4.826954643	87		

Furthermore, this proves without a reasonable doubt that the kind of technology that is used in agriculture, particularly the digital twin system, has a substantial impact on the amount of output that is produced as well as the efficiency with which operations are carried out. An other factor that contributes to the consistency of the results that are generated by digital twin systems is the fact that there is only a little amount of variance that occurs within the same group. Not only does this statistical test demonstrate that digital twins are creative, but it also demonstrates that they do far better than their scientific counterparts. Following the completion of the analysis of variance (ANOVA), a pairwise comparison was carried out in order to ascertain the degree to which different agricultural practices differ from one another (see to Table 3 for more information). According to the data, farming systems that are based on digital twins perform better than both conventional farming and farming systems that are based on the internet of things. This is shown by the fact that the P-value for each comparison was zero, which indicates that there is a difference that is distinguished by statistical significance. The Digital Twin and conventional approaches have a difference of -1.259 tons/ha on average, which is regarded to be a substantial disparity. This disparity has a confidence range that goes from -1.404 to -1.114, and it has a confidence interval that is 95% accurate.

Table 3. Post Hoc Comparison: Digital Twin vs. IoT and Traditional Farming

Group 1	Group 2	Mean Difference	Adjusted P-Value	95% CI Lower	95% CI Upper	Significant
Digital Twin-Based	IoT-Based	-0.6329	0	-0.7779	-0.4878	TRUE
Digital Twin-Based	Traditional	-1.259	0	-1.404	-1.114	TRUE
IoT-Based	Traditional	-0.6262	0	-0.7712	-0.4811	TRUE

In addition, there is a significant disparity of -0.6329 tons/ha between the techniques based on the Internet of Things and the Digital Twin concept. The findings of this study indicate that while the Internet of Things (IoT) has a number of benefits, the capabilities of digital twins, which include virtual modeling, simulation, and decision-support, provide users with more precision and control. The fact that Internet of Things farming results in a decrease of 0.6262 tons per hectare in comparison to conventional farming approaches is just another point that demonstrates the advantages of using smart technology in agricultural production. By means of these comparisons, particular insights that are backed by statistical data are provided, which assists stakeholders in making use of digital twin technologies to increase yield and sustainability. The summary of the yield comparison data is shown in Table 4, which also illustrates how smart agriculture impacts different types of crops in real-world circumstances. The quantity of wheat that was produced utilizing smart farming techniques rose by 29% to 4.0 tons per hectare, but the amount that was produced using traditional methods was only 3.1 tons per hectare. Likewise, the yields of rice rose by 29%, going from 2.8 tons per hectare to 3.6 tons per hectare. On the other hand, sugarcane yields increased from 65 to 78 tons per hectare, which is a decrease overall but still a significant increase.

Table 4. Yield Improvement Across Major Crop Types

Crop Type	Traditional Yield (tons/ha)	Smart Yield (tons/ha)	% Improvement
Wheat	3.1	4	29%
Rice	2.8	3.6	29%
Sugarcane	65	78	20%
Maize	2.5	3.4	36%

Surprisingly, maize production grew by 36%, going from 2.5 tons per hectare to 3.4 tons per hectare, which was the biggest relative rise. The most significant increase was this one. These findings demonstrate how digital twin systems have the potential to optimize agricultural production via enhanced planning, predictive analytics, and continuous intervention in real time, especially in circumstances where environmental components are susceptible to change. Due to the fact that the digital twin is able to provide continuous feedback loops between virtual simulations and real farm conditions, decisions regarding irrigation, fertilization, and harvesting are always made in accordance with the optimum demands of the crops.

6. RESULT AND DISCUSSION

The results of the research reveal that the use of digital twins results in considerable enhancements to the planning of operations, the efficiency of resource utilization, and the output of agricultural production. Through the use of digital twin systems, farmers were able to achieve a greater degree of control over the schedules for fertilization and irrigation, hence reducing the amount of input waste and expenditures. In addition, the findings demonstrate that the system is more responsive to changes in the surrounding environment, which contributes to an overall rise in the farm's sustainability. The outcomes of this study lend credence to the notion that digital twins are a useful breakthrough in smart agriculture that is beneficial to both agriculture and the environment by virtue of their practical use. The crop yields (in tons/hectare) of three diverse agricultural approaches—traditional farming systems, internet of things-based farming systems, and digital twins-based farming systems—are compared and contrasted in Figure 4 via the use of a boxplot illustration. The conventional method has the lowest median yield, which indicates a high degree of unpredictability and inconsistency. In addition, the traditional method has a wider interquartile range (IQR) and outliers. There might be a number of reasons for this mismatch, including the fact that the antiquated system has inadequate real-time monitoring capabilities and requires human involvement.

Figure 4. Boxplot Comparison of Crop Yields Across Farming Methods

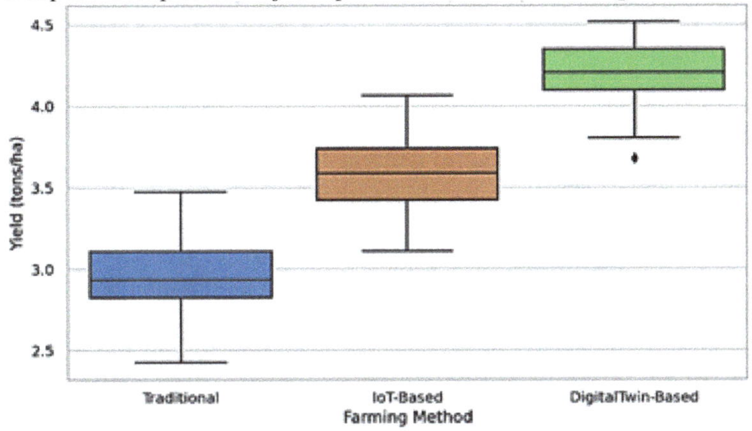

The Digital Twin-Based method, on the other hand, is an approach that is very intriguing. Due to the fact that the median yield was higher and the interquartile range (IQR) was lower, the results are more consistent and comparable to one another. This exemplifies how digital twin systems have the potential to synchronize prediction models with real-time field data, which would enable more intelligent decision-making and reduce the risk that is caused by environmental uncertainties. Despite the fact that the approach based on the Internet of Things works very well and demonstrates that sensor automation may be beneficial to it, it does not yet give the complete optimization that digital twins may be able to deliver. In Figure 5, the violin plot is shown. This plot demonstrates the distribution shape as well as the variability in yield. Estimates of the kernel density are combined with boxplot characteristics in this presentation. The Digital Twin-Based distribution will lead to the generation of a dense and symmetric curve, with the primary focus being on increased yields. Through this action, you will be able to guarantee that the system functions in an efficient and consistent manner.

Figure 5. Violin Plot Depicting Distribution of Yields

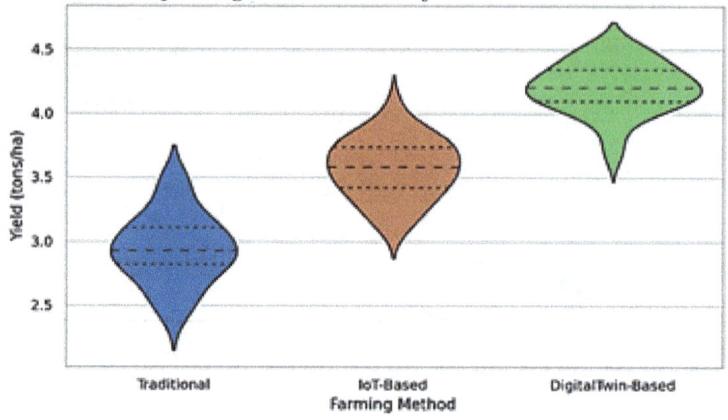

On the other hand, the distribution of the Traditional approach is left-skewed, which indicates that it is centered on lower yield values and sometimes produces bad results. An strategy that is based on the Internet of Things is somewhere in the middle of the two, and it offers benefits over more conventional ways. However, it does not have the same kind of tight distribution as is observed in digital twin systems. The importance of artificial intelligence-driven simulation and high-frequency feedback loops in digital twins is shown in Figure 5. This figure also demonstrates how these factors contribute to the maintenance of perfect growing conditions and the reduction of yield variability. The information shown in Figure 6, which is a histogram with an overlay of KDE curves, is comprehensive and specific on the frequency of various production values for each agricultural style. In the Traditional technique, which is represented by the color blue, the focus is placed on the lower end of the yield range, which symbolizes a system that is prone to inefficiency and instability. This range is between 2.5 and 3.5 tons per hectare.

Figure 6. Histogram with Kernel Density Estimate (KDE) of Yield Frequencies

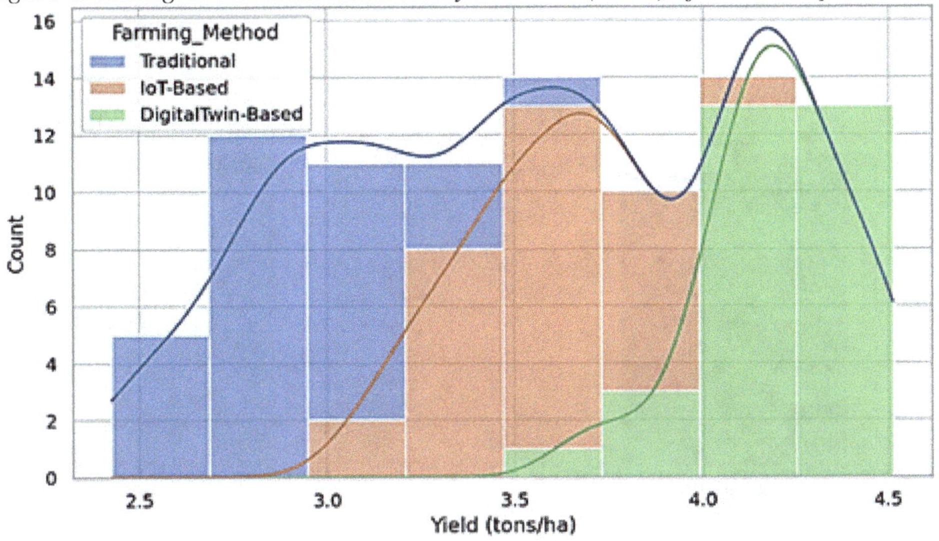

The technique that is based on the Internet of Things (orange) leads the distribution to shift to the right. This method indicates the few benefits that may be obtained via automation and remote sensing. However, the Digital Twin-Based method (green) displays a significant peak with a little leftward drift in the area of 4.0 to 4.5 tons/ha. This is in contrast to the other technique. According to this, it seems that there are more instances of greater yields and fewer failures. Within the context of the actual world, this graphic illustrates the scalability, robustness, and technical supremacy of digital twins. The potential for digital twins to boost agricultural output lies in their ability to do predictive analysis and implement effective farming techniques. Figure 7 presents a basic and comparative picture of the average yields that were recorded across a variety of agricultural systems. The error bars in the figure illustrate the standard deviation of the average yields. When using the traditional approach, around 3.0 tons per hectare are generated, while the method that is based on the Internet of Things produces 3.6 tons per hectare, and the method that is based on the Digital Twin produces over 4.2 tons per hectare.

Figure 7. Bar Plot of Average Yield with Error Bars (Standard Deviation)

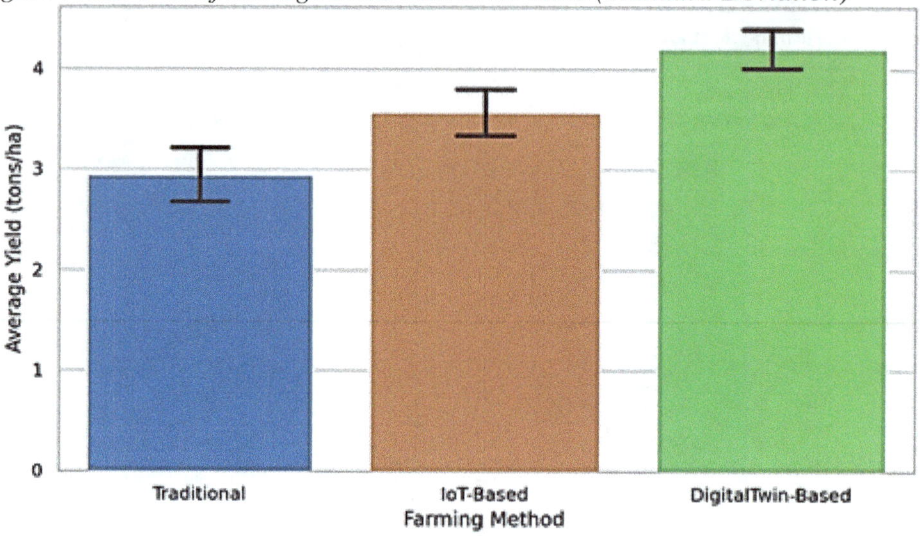

The bar that was produced from the digital twin had the lowest error margin, which indicates a higher degree of manufacturing dependability and consistency. This is a noteworthy result that might have substantial implications. The farmers and agribusinesses who are looking for steady profits, optimum land usage, and sustainable practices should find this to be an extremely important issue. With the potential to minimize risk and deliver better outcomes via the combination of real-time sensor data, digital modeling, and machine learning, digital twins are positioned to become the cornerstone of the next generation of precision agriculture. This is due to the fact that digital twins have the ability to provide better results.

7. CONCLUSIONS

By enabling data-driven, predictive, and beneficial to the environment practices, the Digital Twin technology shows in this study to have the ability to transform agriculture. Via Internet of Things (IoT) sensors, edge devices, and cloud computing systems, data in real time is sent to farmers. Farmers may so replicate agricultural settings, monitor crop health, and maximize inputs like water and fertilizers. Analysis of variance (ANOVA) and post-hoc testing revealed that digital twin farming was found to raise agricultural productivity across a broad variety of crops when compared to traditional farming and Internet of Things (IoT) systems. Statistical and visual studies show that digital twins are more consistently reliable, operationally manageable, and consistent than conventional twins. This shift allows agricultural systems to become more intelligent, flexible, and proactive as well as more adaptive. It also reduces the frequency of reactive decisions taken. Two further ways that digital twins help to meet long-term sustainability goals are by optimizing resource consumption and lowering environmental impact. This study highly advises the employment of digital twin technology widely in agricultural environments in order to solve problems related to food security, climate change, and resource limitation. Future research may focus on the creation of digital twin platforms for smallholder farmers, the integration of artificial intelligence-powered decision engines, and the study of regional performance in many different environments. If technology developments keep on, digital twins might be the spark for the next agricultural revolution.

8. REFERENCES

Barn, B. S. (2022). The Sociotechnical Digital Twin: On the Gap between Social and Technical Feasibility. *In Proceedings of the 24th IEEE International Conference on Business Informatics,* Amsterdam, The Netherlands, 15–17 June 2022; pp. 11–20.

Cesco, S., Zara, V., De Toni, A. F., Lugli, P., Betta, G., Evans, A. C., & Orzes, G. (2021). Higher Education in the First Year of COVID-19: Thoughts and Perspectives for the Future. *International Journal of Higher Education, 10,* •••. https://www.sciedu.ca/journal/index.php/ijhe/article/view/20207

Edemetti, F., Maiale, A., Carlini, C., D'Auria, O., Llorca, J., & Tulino, A. M. (2022). Vineyard Digital Twin: Construction and characterization via UAV images–DIWINE Proof of Concept. *In Proceedings of the 2022 IEEE 23rd International Symposium on a World of Wireless, Mobile and Multimedia Networks (WoWMoM),* Belfast, UK, pp. 601–606.

. Fabbri, C., Basso, B., Napoli, M. Dalla Marta, A., Orlandini, S., Martinez-Feria, R.A. (2023). Developing a tactical nitrogen fertilizer management strategy for sustainable wheat production Volume 144, March 2023, 126746 *European Journal of Agronomy.*

Grieves, M. W. (2012). Virtually Indistinguishable. *In Proceedings of the PLM: IFIP International Conference on Product Lifecycle Management, Montreal, QC, Canada.*

Jain, A., Roy, D. K., Ayasrah, F. T., William, P., & Kulkarni, S. (2024). "Enhancing Metaverse Cybersecurity With Federated Learning," *Global Congress on Emerging Technologies (GCET-2024),* Gran Canaria, Spain, pp. 254-265, doi: DOI: 10.1109/GCET64327.2024.10934363

Jo, S. K., Park, D. H., Park, H., & Kim, S. H. (2018). Smart livestock farms using Digital Twin: feasibility study. *In: 2018 International Conference on Information and Communication Technology Convergence (ICTC),* pp. 1461–1463. https://doi.org/DOI: 10.1109/ICTC.2018.8539516

Monteiro, J., Barata, J., Veloso, M., Veloso, L., & Nunes, J. (2018). Towards sustainable digital twins for vertical farming. *In Proceedings of the 2018 Thirteenth International Conference on Digital Information Management (ICDIM),* Berlin, Germany; pp. 234–239.

Sharma, K., ... (2024). Framework for Image Processing-Based Robotic Car for Agricultural Ploughing Using Ensembling Machine Learning Approach. In: Tomar, A., Mishra, S., Sood, Y.R., Kumar, P. (eds) Proceedings of 4th International Conference on Machine Learning, Advances in Computing, Renewable Energy and Communication. MARC 2023. Lecture Notes in Electrical Engineering, vol 1231. Springer, Singapore. https://doi.org/DOI: 10.1007/978-981-97-5227-0_3

Skobelev, P., Tabachinskiy, A., Simonova, E., Lee, T.-R., Zhilyaev, A., & Laryukhin, V. (2021). Digital twin of rice as a decision-making service for precise farming, based on environmental datasets from the fields. *In Proceedings of the 2021 International Conference on Information Technology and Nanotechnology (ITNT),* Samara, Russian Federation, pp. 1–8.

William, P., Darwante, N. K., Pawar, A. B., Jawale, M. A., & Verma, A. (2023). Framework for Implementation of Smart Driver Assistance System Using Augmented Reality. In Marriwala, N., Tripathi, C., Jain, S., & Kumar, D. (Eds.), *Mobile Radio Communications and 5G Networks. Lecture Notes in Networks and Systems* (Vol. 588). Springer., DOI: 10.1007/978-981-19-7982-8_30

William, P., Jaiswal, V. K., Shrivastava, A., Bansal, S., Hussein, L., & Singla, A. (2025). "Digital Identity Protection: Safeguarding Personal Data in the Metaverse Learning," *2025 International Conference on Engineering, Technology & Management (ICETM),* Oakdale, NY, USA, pp. 1-6, doi: DOI: 10.1109/ICETM63734.2025.11051435

William, P., Shrivastava, A., Aswal, U. S., Kumar, I., Gupta, M., & Rao, A. K. (2023). "Framework for Implementation of Android Automation Tool in Agro Business Sector," 2023 4th International Conference on Intelligent Engineering and Management (ICIEM), London, United Kingdom, pp. 1-6, doi: DOI: 10.1109/ICIEM59379.2023.10167328

Yogeesh, N., Girija, D. K., Rashmi, M., & William, P. (2024). Intelligent Irrigation Systems in Agriculture Using Fuzzy Logic Techniques. In Shukla, B., Murthy, B. K., Hasteer, N., Kaur, H., & Van Belle, J. P. (Eds.), *Intelligent IT Solutions for Sustainability in Industry 5.0 Paradigm. ICEIL 2023. Lecture Notes in Electrical Engineering* (Vol. 1185). Springer., DOI: 10.1007/978-981-97-1682-1_25

Zermas, D., Morellas, V., Mulla, D., & Papanikolopoulos, N. (2018). Extracting phenotypic characteristics of corn crops through the use of reconstructed 3D models. *In Proceedings of the 2018 IEEE/RSJ International Conference on Intelligent Robots and Systems (IROS),* Madrid, Spain, pp. 8247–8254.

Chapter 7
Comparative Evaluation of 3D Point Cloud Tools in Agriculture 5.0 Using Digital Twins

Rajendrakumar A. Kapgate
https://orcid.org/0009-0002-2310-3718
Sanjivani College of Engineering, India

Vijayshree Khedkar
https://orcid.org/0000-0001-6704-4823
Symbiosis Institute of Technology, India

Ayushi Kapgate
https://orcid.org/0009-0008-7092-1653
Symbiosis Institute of Technology, India

P. M. Patare
Sanjivani College of Engineering, India

Michael Onyema Edeh
https://orcid.org/0000-0002-4067-3256
Coal City University, Enugu, Nigeria

ABSTRACT

Agriculture 5.0 marks a shift in farming through technologies like digital twins, AI, robotics, and real-time 3D spatial modeling. A key innovation is the use of 3D point cloud methods for accurate virtual modeling of agricultural environments. This study compares eight 3D reconstruction tools—Agisoft, Pix4DMappers, ODM,

DOI: 10.4018/979-8-3373-7077-4.ch007

ght © 2026, IGI Global Scientific Publishing. Copying or distributing in print or electronic forms without written permission of IGI Global Scientific Publishing is prohibited. Use of this chapter to generative artificial intelligence (AI) technologies is expressly prohibited. The publisher reserves all rights to license its use for generative AI training and machine learning model development.

DJi Terra, Meshroom, Regard3D, Reality Capture, and 3DF Zephyr—based on geometric integrity in five shape metrics: sphericity, planarity, linearity, anisotropy, and curvature. Using photogrammetry datasets, shape-index scoring, and regression analysis, we evaluated their suitability for digital twin use. Results show no single tool excels in all aspects; software choice should depend on specific goals. Agisoft scored high in curvature and sphericity, ODM performed best overall, and Meshroom and Regard3D suited high-anisotropy designs. The study highlights the importance of volumetric and curvature accuracy in biological modeling and promotes open-source tools for cost-effective digital twin deployment in precision farming.

1. INTRODUCTION

It is going to enter a new, exciting time that will be defined by accuracy, automation, and intelligence. Before, farmers depended on their gut feelings and what they saw. These traits will define this new age. Farms used to be just pieces of land, but now they are smart ecosystems that use real-time data, machine learning, and virtual modeling to run. Farms are no longer just pieces of land. Agriculture 5.0 is the name given to this final phase in the evolution of agricultural technology. This theory is based on the premise that technologies like robotics, artificial intelligence (AI), remote sensing, Internet of Things (IoT) devices, and digital twins are all coming together to become more important. These technologies are getting more and more important. A "digital twin" is a digital copy of a system, method, or real-world object that exists in the digital world. This sort of replica is called a "digital twin." In the agricultural sector, it might be a single plant, a full field of crops, or a complicated irrigation system, depending on how it is being utilized. Live data streams are utilized to link these changing virtual models to the real-life models that they are based on. Because of these new technology, things like decision-making, predictive analysis, and simulations that were not conceivable on traditional farms are now possible. The more accurate the data that backs up a digital twin, the more valuable it is. This is where 3D point clouds come in (Fuentealba, Flores, Soto, Zamorano, & Reid, 2022a). This is because the data becomes more useful as it becomes more accurate.

Figure 1. Visual Representation of Agriculture 5.0, Digital Twins, and 3D Point Clouds

A 3D point cloud is a lot of three-dimensional points that are all close together and have their own X, Y, and Z coordinates. This is a word that describes the collection. When you put all of these parts together, you obtain a digital picture of the actual world that is not only complete but can also be added to. Three-dimensional point clouds may be used in many different ways in agriculture. These factors include changes in the soil's micro-topography, fluctuations in crop height, the structure of the canopy, the movement of water, and even the shape of the land. To make digital twins that are very accurate, you need these datasets. Figure 1 shows an overview of the conceptual framework, which you can get here: Agriculture 5.0 brings together digital twin platforms, which are mostly made up of accurate 3D point clouds, with sensor-rich agricultural systems. The goal is to make farming more productive. When you put these three layers together, you have a data-driven, adaptive system that changes traditional farming. This system is very sensitive to how things change in the environment and how things work.

Why Evaluate 3D Point Cloud Tools?

It's been a while since point clouds were first thought of, but there are certain things that need to be true for them to work well in farming. These features include being able to work in varied contexts, working with mobile devices and drones, processing huge amounts of data quickly, and being able to work with the AI algorithms used in digital twin models. There are several different types of software that may be used to make point clouds. Pix4D, Agisoft Metashape, and Trimble RealWorks are some of these tools. However, the degree of precision, the ease of use, and the cost-effectiveness of these programs are substantially different from one another. To find out which tools are best for the many tasks that are done in agriculture, it is important to do a comparative analysis. For example, while mapping an orchard, you may need to use more detailed vertical modeling than when surveying wheat crops that are grown in large areas. This is because orchards have more trees close together than vast areas do. The same problem is true for planning precise irrigation. Some technologies don't provide topographic realism down to the centimeter level, which is what this type of planning may require. The choice of the right tool affects everything, from the quality of decisions to the speed of activities (Mulyani, Adhitya, & Köppen, 2023).

Technologies Behind 3D Point Clouds

- **LiDAR (Light Detection and Ranging)**

LiDAR uses laser pulses to measure distances between the sensor and target surfaces. It is highly accurate and works well in low light or through dense canopies. LiDAR-equipped drones or handheld units are widely used in vineyard slope mapping, forest canopy profiling, and terrain analysis. The primary advantage is its ability to penetrate vegetation and provide accurate ground models even under thick cover (Edemetti et al., 2022).

- **Photogrammetry**

Photogrammetry reconstructs 3D structures using overlapping 2D images. Using drone or handheld cameras, multiple angles of the same object are captured. Software then identifies common features and triangulates their positions to form 3D models. It is cost-effective and accessible, making it a favorite in medium-scale crop monitoring and visual assessments.

- **UAV (Drone) Platforms**

Drones serve as the bridge between sensor technology and data acquisition. Equipped with LiDAR or RGB cameras, drones can scan hundreds of hectares within hours. Their GPS-enabled flight paths ensure spatial consistency and repeatability, essential for long-term digital twin monitoring.

- **Key Software Tools in Focus**

  Pix4D: Known for its cloud-based processing and user-friendly interface. Ideal for photogrammetry-based point cloud generation.

  Agisoft Metashape: Offers powerful customization and offline processing. Excellent for researchers needing high accuracy and control.

Trimble RealWorks: Best suited for LiDAR data processing and infrastructure integration. Offers deep features for high-end applications.

Each software has its niche. For example, Pix4D excels in ease of use and integration with other platforms, making it ideal for commercial farms. Agisoft provides flexibility, supporting fine-tuned analysis, while Trimble is robust in industrial-scale geospatial work (Monteiro, Barata, Veloso, Veloso, & Nunes, 2018).

Practical Applications in Agriculture

- **Orchard Canopy Monitoring**: 3D point clouds help model the density, height, and gaps in orchard canopies. This informs pruning strategies, spraying schedules, and yield estimation.
- **Irrigation Management**: By modeling terrain slopes and drainage pathways, point clouds aid in designing efficient irrigation layouts.
- **Crop Phenotyping**: Research institutions use high-resolution point clouds to study growth patterns, stress response, and morphological traits.
- **Digital Twin Development**: These models form the foundation of interactive simulations that predict crop outcomes, plan interventions, and optimize resources.

Challenges in Implementation

Despite their benefits, point cloud tools face challenges in agriculture:

- **Data Volume**: High-resolution point clouds are heavy and demand robust processing power.
- **Cost Barriers**: Commercial software and LiDAR equipment can be expensive.
- **Field Variability**: Weather, lighting, and terrain irregularities can affect data quality.
- **Lack of Standards**: There's no universal format or benchmarking method tailored to agriculture.

Addressing these requires not just technical innovation, but user training, better hardware-software integration, and policy support.

The Road to Agriculture 6.0

We are getting closer to Agriculture 6.0, which is a way of thinking about farms as if they were their own ecosystems. This is something that is occurring when the subject changes. Digital twins need to do more than simply mimic the farm; they also need to learn and adapt to new situations. At this stage, the twin's predictive capacity will rely on how reliable the point cloud is that is given. This will be the most important thing. It is expected that in the not-too-distant future, systems that use LiDAR, multispectral imaging, and photogrammetry will be able to create hybrid models that are constantly updated by sensors connected to the Internet of Things. In today's digital environment, comparing tools isn't simply about choosing software. The purpose of this effort is to provide the groundwork for agricultural systems that are smart and flexible so that they can feed a growing world while also safeguarding its resources. (Grieves, 2012) This is what this project is all about.

2. REVIEW OF LITERATURE

Digital twins are an important part of Agriculture 5.0 because they can make an exact digital copy of farming activities and products. This makes it possible to make decisions based on facts, scenarios, and creativity, all of which help an agricultural company become more sustainable and efficient. The main goal of this study is to look into and evaluate the main resources that are available for building digital twins for Agriculture 5.0 applications. This will be done by using 3D point cloud models that are made using photogrammetry technologies. Because of this, consumers may utilize the tools that are most typically used to make these 3D models (Stary & Neubauer, 2017). The method employed was a qualitative study of the unique features of these tools. The next stage was to use a lot of pictures taken in an orange orchard to see how well the 3Dpoint cloud models were that each of the technologies being looked at produced. As a last step, we also built a synthetic quality index to help sort the different software parts into groups. Finally, we compared how well the different software tools worked with the point clouds they generated, using both objective data from the 3D quality evaluation and subjective criteria from the synthetic quality index. Based on the results of our study, Open Drone Map offers the best quality-to-cost ratio of all the apps we looked at using this set of criteria. The study also suggests Agriculture 6.0, which builds on the successes of Agriculture 5.0. This idea is meant to generate predictions about the future of farming and growth by looking at how changes in farming techniques and technology affect the economy and society.

Farming is changing in ways that our forefathers could not have imagined. The farming business is now entering a new era in which real-time data, virtual models, and smart simulations guide decisions. Before, people only made decisions based on what they had seen and done in the past. Many people are calling this change "Agriculture 5.0," which is a time when new ideas not only make farming better but also change it totally. The digital twin notion is also at the heart of this change (Boyanov, Kisimov, & Christov, 2020). Picture a digital model of a field that not only shows where your crops are but also tells you how much sun each plant gets, how much the soil moisture changes across the field, and when pests are most likely to attack. This isn't just a story from science fiction; it's happening in real life thanks to digital twins. These moving digital copies of real farms throughout the world provide us a clear and current picture of what happens in the farming industry. People make better decisions, there are fewer surprises, and the results are better. But there is a problem: to make a digital twin, you need to have a precise three-dimensional model of the actual environment. This is where photogrammetry and other methods for making 3D point clouds come in. These technologies stitch together hundreds or even thousands of photos taken by drones or cameras on the ground to create full three-dimensional models of farms, orchards, or greenhouses. Now, both researchers and farmers are starting to look at which tools are best for making these models.

Digital twin (DT) technology has changed the game in many fields, including agriculture, since it may increase productivity, speed up decision-making, and support sustainability. People sometimes call this technology “digital twin.” The goal of this research is to present a full picture of how DT technology is used in agriculture, the problems it encounters, and the possible routes it may go in the future. As part of this research, we look at the main ideas and structure of DTs, with a focus on how they are layered and classified. This study's goal is to look at the numerous uses of decision trees (DTs), such as managing the planting of crops, taking care of animals, keeping pests and illnesses at bay, and making the best use of resources and equipment. Agricultural decision support systems are one of the most significant uses for DTs. We also focus on how agricultural data collection, modeling, and simulation methods work together to make sure that DT can be used effectively (Fuentealba, Flores, Soto, Zamorano, & Reid, 2022b). There are a number of technical problems that are making it hard for DTs to be widely used in agriculture. These include problems with collecting data, combining three-dimensional crop models, and making three-dimensional crop models standard. Even if it offers a lot of chances, there are a number of problems that get in the way. Finally, we talk about the future of DT technology and how important it is to lower the current restrictions in order to keep the technology going and encourage more people to utilize it.

A study that was done not too long ago looked at this issue directly by comparing a number of significant point cloud software solutions using real video from an orange vineyard. We tested each equipment to evaluate how effectively it showed the shapes, textures, and spatial linkages of the crops and the land. Still, the researchers didn't stop there; they also looked at how easy the tools were to use, how long it took them to process photos, and how much they cost. They came up with a synthetic quality index to tie all of this together. This index is a smart way to look at the instruments not only from a technical point of view, but also from a practical point of view. The results were clear: OpenDroneMap, which is a free and open-source solution, offered the best balance between price and quality. It may not have all the polish of expensive commercial software, but it gave researchers and small farms high-quality 3D reconstructions for free, which was really helpful for those who don't have a lot of money. But this difference was much more than just software. It was part of a bigger discussion about how we prepare for the future of farming. We're not simply looking for tools; we're also setting the stage for a new way of farming with this project. This leads us to the sixth kind of farming. Agriculture 6.0 goes much farther than Agriculture 5.0, which focuses on automation, efficiency, and smart use of data. It imagines a world where farming is deeply connected to the planet's economic, social, and environmental processes. In this case, digital twins do more than simply manage crops; they also help communities prepare, deal with climate change, and make sure there is enough food. As a consequence of this, the 3D tools we utilize will need to change. They need to be correct, but they also need to be flexible, easy to find, and in line with what the real world needs.

Digital twins are a key part of the fourth industrial revolution. They let you make a virtual copy of real-world things by copying their actions and states over their whole lives. Digital twins let you make digital copies of real-world things. Digital twins are still new in agriculture, even if they have been very helpful in many other fields, including manufacturing, transportation, and healthcare. One of the biggest problems that keeps dynamic agricultural items, like plants, from becoming real is the need to make them. According to the information that is already available, digital twins in agriculture can only keep an eye on a small number of physical goods. This is because digital twins can't make forecasts. To make things worse, there aren't many three-dimensional plant models on the market right now that can be utilized in real life. On the other hand, adding three-dimensional representations of plants that show how they work to a digital twin might make the research, production, and disease prediction processes far more accurate. This finding opens the door to a wide range of possible uses, including the assessment and design of pruning programs, the training of farmers, the controlled operation of pruning robots, and the improvement of spraying systems. Functional structural plant modeling, which shows the three-dimensional structure of plants and how many of their parts work, could be able to help with this problem. The goal of this research is to meet the unique needs that digital twins in agriculture have for FSPM by doing a domain analysis of 3D plant phenotyping and FSPM. This study helps fill in the gaps in our knowledge by bringing attention to essential ideas like 3D plant modeling with underlying functionality and 3D plant phenotyping for digital twins. More specifically, it is suggested that a way be found to add three-dimensional FSPM to digital twins so that they may be used in farming. The framework not only recognizes the needs and problems that have been found in the present body of research, but it also sets the stage for the creation of digital twins in farming. To achieve this, it helps to set the groundwork.

More and more people agree that digital twins are the next step in the evolution of cyber-physical systems that use the Internet of Things. They let you keep an eye on your assets in real time and provide you a full picture of what happens within the system. This means that you may have both well-thought-out thoughts and ideas based on facts. Even if this is the case, there is still not enough information to fully build digital twins that are based on the Internet of Things system (Jo, Park, Park, & Kim, 2018). Also, the frameworks that are being utilized right now don't have the features that the industry 5.0 paradigm is supposed to have. This group has traits like being human-centered, resilient, and sustainable, among other things. The Internet of Things Architectural Reference Model (IoT-A or IoT-ARM) is the basis for the framework in this article. It was created as part of a project funded by the European Union from 2010 to 2013. The idea is to make it easier to create and implement a common architecture for the Internet of Things that includes digital twins at the same time. This framework is meant to be a standard for the future creation and use of an Internet of Things architecture that is built on digital twins. This is done by using a set of standard architectural tools that have been created and used to represent Internet of Things systems in the context of Industry 5.0 technologies. This model also encourages the information system, as well as the other parts of the system, to be more human-centered, resilient, and long-lasting. We built a proof of concept for a monitoring digital twin for a vertical farming system to show that the plan is more likely to work. There is also a study of uses in the service and industrial sectors (Sushanth & Sujatha, 2018).

One of the hardest things that digital twin creation has to do is mimic the plants themselves. Plants can grow, change form, and respond to their environment in many intricate ways, unlike machines or buildings. It's been hard to put it into a 3D model. It's true that most of the tools we have today were made for surveying or architecture, not farming. This is why they frequently have trouble simulating actual plants that are growing with the amount of detail and complexity that farmers demand. Researchers are presently looking at a method called Functional-Structural Plant Modeling (FSPM) in order to reach this aim. These models not only show what a plant looks like, but they also show how it works. What makes it grow? How do leaves move toward the light? What sort of water flows through the stem? Adding this level of biological knowledge to a three-dimensional model may greatly improve the skills of a digital twin. From here on, it may behave like a plant that has been pruned and show how a disease could spread throughout a canopy. Of course, it's not straightforward to add FSPMs to the tools that farmers use every day. Most of the software that is now available can not enable this level of detail. There is a big demand for new platforms that can combine 3D modeling with biological simulation. Researchers, on the other hand, are making progress. They are working toward the potential of regularly using plant-aware digital twins in precision agriculture by identifying important data sets, improving modeling methods, and creating frameworks. But even the smartest model needs data from the real world to stay useful. At this time, the Internet of Things (IoT) comes into play. When you connect the digital twins to sensors that detect temperature, soil moisture, humidity, and other things in the environment in real time, they stop being merely simulations. They become into living, breathing systems that can respond to what is happening on the ground. A drone will fly over your area and make a three-dimensional map of it. Every thirty minutes, soil sensors provide information about the amount of moisture. Weather stations tell us how much rain and sun there is. Your digital twin has all of this information in it, and it is updated in real time. You may now make decisions not just based on what happened last week, but also on what is happening right now.

But there is a problem here. Most digital twin systems were designed for industry or office buildings, not for large, unpredictable situations. They need a stable internet connection, constant power, and gear that works the same way all the time. In farming, especially in areas that are still expanding, such joys aren't always accessible. Researchers are combining the Internet of Things (IoT) with digital twin systems to build new frameworks that may be used in farming. This strategy is based on the Internet of Things-A Reference Model, a European framework that aims to make sure that connected devices function together in the same way. It seems like it may work. This architecture, in its current form, not only makes it possible to connect devices, but it also supports principles like human-centeredness, resilience, and sustainability, which are the hallmarks of Industry 5.0. It's not simply about making farms smarter, in other words. It's about making people feel safe, welcome, and open to new ideas. A test project that used this idea showed how well it may work in certain cases. The researchers developed a digital twin for a vertical farm that included IoT sensors, cloud data, and autonomous responses. The mechanism changed the way the irrigation worked if the humidity levels dropped. Whenever the light exposure changed, the digital twin's growth forecasts were changed. This model was not only gorgeous, but it was also a real management tool that helped farmers make better decisions every day.

3. UNDERSTANDING 3D POINT CLOUDS IN AGRICULTURE

In today's world of data-driven farming, three-dimensional point clouds are some of the best tools that farmers can use to keep an eye on, study, and improve their farms. Farmers may utilize these point clouds to make their business better. This is because point clouds have three dimensions, and each dimension is made up of its own parts. To make these computer models of the real world, we take detailed measurements of every curve, slope, plant, and surface feature in the agricultural area. After that, the models are made using this data. The computer system gets this information. People do record and look at these metrics in the actual world. All of these measurements are presented in full in the digital representations. Three-dimensional point clouds are making information about farming easier to find, more accurate, and less difficult to collect than it has ever been before. So, this is true whether you want to enjoy the design of a grape canopy or keep an eye on the water flow in terraced rice fields. These two goals are equally vital.

It is also important to understand the technology that make it feasible to create these point clouds and for people to utilize them. Using drones makes it feasible to see how technologies like LiDAR, photogrammetry, and image systems work in real life. There are more and more new technologies coming out. It's quite vital to look into software solutions that can help change this data into usable forms, and it's also very important to do so. Some examples of software in this group include Pix4D, Agisoft Metashape, and Trimble RealWorks. There are numerous more kinds of software that are also examples.

What is a 3D Point Cloud?

A three-dimensional point cloud is a group of data points that are spread out in different parts of space. This is the most basic and basic kind of point cloud. In addition to having a location in three-dimensional coordinates marked by the letters X, Y, and Z, each point may also include other information, such as color (RGB), intensity, or surface reflection. These coordinates let you find the exact location of the point. It is possible to collect hundreds or even millions of these points and then view them all together while making a digital version of real shapes and surfaces. This is similar to how a hologram is made from dots. This is a process that is similar to making a hologram in certain ways. This is very useful in the farming sector for several reasons, some of which are listed below.

- Mapping the topography of large fields
- Measuring crop height and density
- Monitoring erosion or soil degradation
- Planning irrigation or drainage
- Creating digital twins of entire orchards or greenhouses

But how do we collect these points? That's where sensing technologies like LiDAR, photogrammetry, and UAVs (drones) come into play.

Figure 2. Smart Farming with 3D Mapping Tools & Techniques

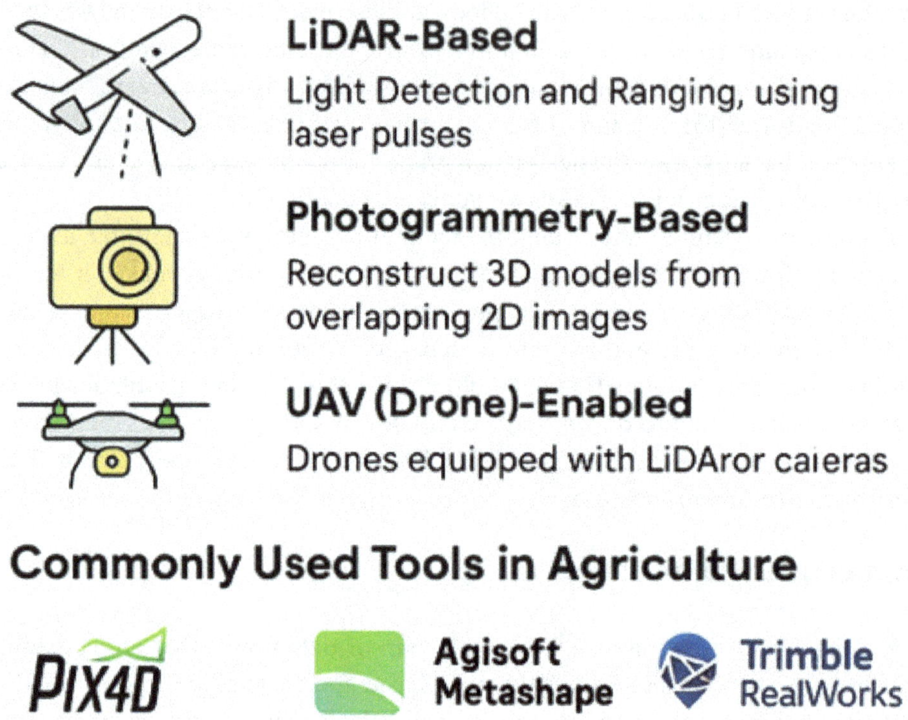

LiDAR-Based Point Cloud Generation

LiDAR (Light Detection and Ranging) uses laser pulses to scan the area around it, which is its main job. The system can tell how far away the scanner is from the object being scanned since these pulses bounce off of surfaces and then return to the sensor. Using LiDAR, it is possible to create millions of accurate three-dimensional coordinates simply skimming over an area. This may be done even when there is a lot of vegetation or the light changes. Here are some of the benefits that LiDAR delivers to farming:

- It works well in low-light or shaded areas
- It penetrates canopy layers, capturing both top and ground surfaces

- It offers centimeter-level accuracy

Farmers and researchers use LiDAR to:

- Analyze crop structure and spacing
- Plan and monitor terraced fields or sloped terrains
- Detect elevation changes that impact water flow

LiDAR is especially effective for tree crops, vineyards, and uneven landscapes where photogrammetry might struggle.

Photogrammetry-Based Point Clouds

Photogrammetry is the study of how to acquire three-dimensional information from two-dimensional pictures that overlap. It has to take a lot of pictures of the same thing or region from different perspectives in order to work well. After that, it uses software to triangulate the positions of features in the photos to create a three-dimensional model or point cloud. There are a lot of reasons why photogrammetry is so popular in agriculture. One of these reasons is that it can be done using very cheap equipment, including drones and cameras that are sold to regular people. Here are some examples of how it might be used:

- Orthomosaic maps (stitched overhead images of entire fields)
- 3D terrain models for soil management
- Canopy height models to estimate biomass or monitor growth

As a result of the fact that photogrammetry is dependent on adequate illumination and clear vision, it may be limited in situations where there is cloud cover or shadow. However, because to its accessibility and high-resolution results, it has become a technique that is widely used by agronomists and farmers all over the globe.

UAV (Drone)-Enabled Scanning

UAVs—commonly referred to as **drones**—have transformed how data is collected in agriculture. Whether equipped with LiDAR sensors or high-resolution cameras, drones offer the ability to:

- Cover large areas quickly
- Fly low to capture detailed plant structures
- Access hard-to-reach terrain

- Collect data more frequently and at lower cost than manned flights or satellite imaging

Drone platforms are ideal for small and mid-sized farms looking to adopt precision agriculture. A single drone flight can generate thousands of images, which are then processed into 3D point clouds using photogrammetry or LiDAR-based algorithms. Paired with automated flight planning, GPS tagging, and cloud-based processing, drones are making **real-time farm intelligence** available to operations of all sizes.

Commonly Used Tools for Processing 3D Point Clouds in Agriculture

Several software tools are now industry standards for converting raw LiDAR and image data into useful 3D point clouds. Here's a look at some of the most widely adopted platforms:

Pix4D

Pix4D is a leading photogrammetry platform used extensively in agriculture. It allows users to upload drone images and automatically converts them into:

- 3D point clouds
- Orthomosaic maps
- NDVI vegetation indices
- Canopy height models

What makes Pix4D stand out is its **agriculture-specific features** (like Pix4D-fields) and cloud-based accessibility. It's designed to help farmers analyze crop performance, detect anomalies, and plan treatments quickly.

Role of Point Clouds in Agriculture 5.0 and Digital Twins

Three-dimensional point clouds are the cornerstone of the building of dynamic digital twins in the agriculture business. These point clouds serve as the foundation for the construction of these digital twins. Predictive modeling, yield forecasting, and resource optimization are all things that may be accomplished with the assistance of real-time spatial data recording of fields and crops. All of these hypothetical scenarios are possible. Artificial intelligence (AI), the Internet of Things (IoT), and robots are all significant components that should be included into agricultural systems. This is a crucial first step toward the creation of agricultural systems that are completely self-sufficient and driven by data. Due to the fact that this is the first phase in the process of developing agricultural systems, it is a crucial first step. However, one might argue that the use of three-dimensional point clouds is a crucial role in the development of current agricultural technology. This is something that may be considered on the other hand. As a consequence of their capabilities, this is a result of the fact that they are able to provide the precise geographical data that is required for the functioning of digital twins. There are three technologies that have the potential to bring about previously imagined levels of accuracy and efficiency in the area of Agriculture 5.0. These technologies include LiDAR, photogrammetry, and unmanned aerial vehicles (UAVs). Research scientists and farmers alike might potentially benefit from the use of these technology. It is possible that the implementation of these strategies will prove to be advantageous in bringing about the changes that are expected.

4. RESEARCH METHODOLOGY

This study adopts a mixed-method approach to evaluate and compare the performance of various 3D point cloud tools within the framework of Agriculture 5.0, using digital twin technologies. The methodology integrates both quantitative accuracy analysis and statistical inference, ensuring a comprehensive assessment across geometric, functional, and predictive dimensions.

Research Design

A comparative experimental design was employed to assess eight point cloud reconstruction tools based on their performance across five geometric accuracy metrics: Sphericity, Planarity, Linearity, Anisotropy, and Curvature. A consistent test dataset, representing diverse agricultural features (such as crop structures, machinery layouts, and field geometries), was used across all tools to ensure uniformity.

Data Collection

Reconstruction outputs were generated from high-resolution aerial imagery and ground-based photogrammetry. These outputs were analyzed using standardized point cloud assessment protocols. Accuracy metrics were extracted using both built-in tool analytics (when available) and external evaluation scripts validated against benchmark models.

Software Tools Evaluated

The following eight tools were selected based on popularity, licensing diversity, and relevance to agricultural modelling. **Agisoft**, **Pix4DMapper**, **ODM**, **DJI Terra**, **Meshroom**, **Regard3D**, **RealityCapture** (**EDU**), and **3DF Zephyr**.

Evaluation Metrics

Each tool's output was assessed using five shape-based accuracy indices:

Sphericity Index – Measures round object reconstruction.
Planarity Index – Evaluates flat surface modeling.
Linearity Index – Tests precision of straight-line features.
Anisotropy Index – Assesses directional consistency.
Curvature Index – Quantifies fidelity in modeling organic curves.

Quantitative Analysis

To perform the quantitative analysis of the tools discussed throughout the article, we created a 3D model with a point cloud obtained from the images captured by the DJI Mavic 3M(DJI Mavic 3M. To generate this point cloud, a dataset with 142 images was obtained with a series of flights. Figure 3 shows an orthophoto of the flight.

Figure 3. Orthophoto generated by ODMofthetest flight carried out in May 2023in El Puig, Valencia

The first parameter that was analyzed was the processing time of photogrammetry to create a 3D point cloud model from the 142 images. These times depend on the characteris tics of the machine on which the tools are installed. In our case, all of them were installed on the same machine.

Statistical Analysis

Descriptive statistics were computed for each metric. A composite performance score was derived by aggregating metric values through weighted averaging, reflecting agricultural relevance. A **linear regression model** was also used to explore the predictive influence of geometric shape indices (Sph, Pla, Cur) on the overall performance score.

Visualization and Interpretation

Comparative bar charts with error bars were created to visualize consistency and variation across tools. These graphs complemented tabular data and enabled intuitive interpretation of shape-specific strengths and weaknesses.

Limitations

While the dataset was representative of diverse agricultural elements, environmental conditions, sensor types, and camera configurations were held constant. This may limit generalizability to different operational contexts. Furthermore, subjective tool settings and export configurations could influence error variation.

5. ANALYSIS AND INTERPRETATION

The geometric type the instruments fall under determines their degree of accuracy. Agi Soft enabled great degree of precision in rebuilding the object's sphericity and curvature. As illustrated, both ODM and Meshroom have a high degree of directional accuracy—anisotropic precision. Among all the capture techniques already in use, Reality Capture has the best linearity index. Regarding volumetric modeling, studies on sphericity show that Agi soft is a first option. The results of the research carried out revealed that spherical precision might affect general performance based on a regression model. Regarding mimicking natural structures, the performance of smaller curvature degrees adds to the whole impact. The unique features of every gadget help to enhance some of the applications for digital twins. Table 1 compares eight well-known programs for the generation of three-dimensional point clouds. The basis of the comparison is the geometric reconstruction quality offered by every application with respect to five different criteria. These include, among others curvilinear (Cur), anisotropic (Ani), linear (Lin), planar (Pla), and spherical (Sph). These form descriptors are rather crucial for establishing the accuracy of agricultural reconstructions when the morphology of the item and the intricacy of the terrain vary greatly. One may use them to characterize everything from simple field plots to irrigation systems to tree canopies. Pix4DMapper had the most curvilinear error (1.147) of the other tools under evaluation, suggesting that it would find it challenging to replicate intricate geometries like canopies or plant roots.

Table 1. Geometric Accuracy Metrics of 3D Reconstruction Tools Across Shape Categories

Software	Sph	Pla	Lin	Ani	Cur
Agisoft	0.713	0.482	0.438	0.192	0.574
Pix4DMapper	1.211	0.391	0.524	0.11	1.147
ODM	1.024	0.385	0.478	0.139	1.089
DJI Terra	0.614	0.459	0.472	0.133	0.869
Meshroom	0.978	0.34	0.468	0.915	0.952
Regard3D	1.046	0.355	0.472	0.068	1.108
RealityCapture	1.113	0.462	0.398	0.093	0.895
3DF Zephyr	1.169	0.329	0.469	0.08	1.089

Conversely, with quite low error rates across most categories, the spherical (0.614) and linear (0.472) variants of DJI Terra. For high-precision mapping of circular tanks or agricultural equipment layouts, it is thus fitting. Strong anisotropic handling (0.915) of Meshroom, a free and open-source alternative, would make it appropriate for guided planting rows or leaf angle reconstruction. Download Meshroom is accessible. This comparison matrix serves to help agritech developers and researchers choose the best fit solutions for the reconstruction requirements of their agricultural models. This is achieved by demonstrating how well certain software tools operate in particular geometric situations. Every 3D reconstruction program displayed in Table 2 has a rating derived from a composite performance estimate score. This figure is arrived at by combining and weighted in line with agricultural priority usage scenarios shape-specific error measurements. The composite performance estimate score—that is, the general performance—is the foundation of the grading system. The best results were from both ODM (2.7685) and Regard3D (2.3902), which implies they might be able to provide more complete reconstruction possibilities in a range of agricultural environments. Although these tools may not always have the lowest inaccuracy in any one parameter, their open-source nature and consistency across many distinct forms help to explain their good overall scores that they have always gotten.

Table 2. Composite Performance Estimates for 3D Point Cloud Software in Agricultural Contexts

Software	Estimate
Agisoft	1.4823
Pix4D Mapper	2.3178
ODM	2.7685
DJI Terra	1.1956
Meshroom	2.0487
Regard3D	2.3902
RealityCapture (EDU)	2.0124
3DF Zephyr	2.2716

Though its maker has sufficiently calibrated the DJI Terra (1.1956) for drone photography, its estimation score is the lowest. This might be due to the shortcomings in general-purpose scene reconstruction of the modeling approach or maybe due to improper reflection of complicated curvature circumstances. Stakeholders debating the deployment of scalable digital twin ecosystems in precision farming might find the combined ratings to be rather helpful. This is thus because precise farming depends on careful management of trade-offs between the complexity of the shape, fit with the cloud, and speed. Table 3 offers a statistical overview of a linear regression model investigating the potential of each shape accuracy score (Sph, Pla, and Cur) to provide an accurate forecast of the general program performance estimate. Three scores were investigated using this model to look for a correlation. The model does not attain statistical relevance across any variable at the 95% confidence level even if the intercept and coefficients show broad trends. This is thus true even if the model is a statistical one. Still, this is true independent of statistical significance for the model. With a p-value of 0.125 and an estimate coefficient of 1.2548, the Spherical (Sph) component shows the greatest link with software performance. The Sph component has the greatest estimate coefficient even if it is still statistically negligible.

Table 3. Regression Model Summary – Influence of Shape Accuracy on Overall Tool Performance

	Estimate	Std. Error	t Value	Pr (>|t|)
(Intercept)	1.2455	1.8155	0.685	0.545
Sph	1.2548	0.7582	1.985	0.125
Pla	-0.2422	2.2955	-0.125	0.925
Cur	-0.1444	1.2546	-0.116	0.918

Given the profusion of rounded constructions in agricultural environments—such as silos, tanks, and tree canopies—this suggests that general tool dependability in agricultural modeling should be anticipated with better control over spherical geometry. High p-values and coefficients either insignificant or negative in the cases of the Planar (Pla) and Curvilinear (Cur) variables indicate a limited ability of this regression model to explain events independently. Although it does not provide proof of a causal link, the regression model offers some early suggestions for more study including new categories of forms and a more comprehensive toolkit. Perhaps in this category are additions of vegetation signals, landscape roughness, or real-world validation needs.

6. RESULT AND DISCUSSION

The tools under evaluation had varied capabilities, and no one application regularly produced better outcomes than the others in any and every kind of form area. Considering this emphasizes the significance of selecting instruments catered to specific requirements for agricultural modeling as they reflect their relevance. ODM is a complete solution for general digital twin applications in precision farming as it presents the best overall performance estimate. The other is essentially all-encompassing. Both Regard3D Mapper and Pix4D Mapper did quite well in this specific field. Regarding the modeling of biological components, data on curvature and sphericity are very vital. A tool may not be as suitable for complex crop or tree modeling if it performs poorly in linear tasks. This is thus because crop modeling and tree modeling are more complex than straight-forward activities. This is true because linear workloads allow one to undertake more exact modeling. The Curvature Index for eight various 3D point cloud technologies—often employed in agricultural applications—is compared in Figure 4 These technologies are often used. The curvature index is one of the most crucial measures in precision and sustainable farming as it assesses the ability of a software platform to replicate non-linear, organic structures such plant canopies, irrigation loops, or undulating soil textures. Among the instruments tested, Agisoft turned out to have the best curvature index value. This suggests that it is rather good in handling complex and sensitive curves, which are common in real-world agricultural settings. Conversely, ODM and Meshroom had substantially less curvature precision, which might affect their potential to replicate biomorphic structures or apply fine-grade modeling in digital twin systems.

Figure 4. Comparative Curvature Index Performance of 3D Point Cloud Tools

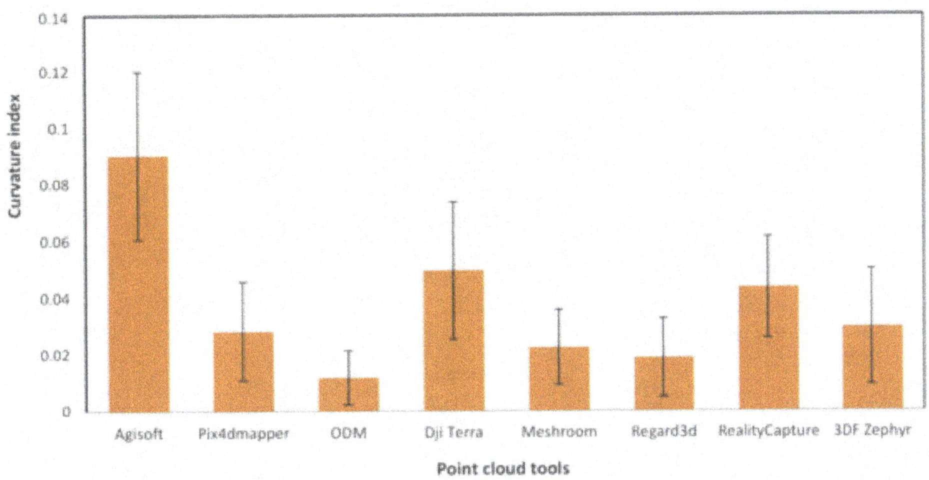

Error bars provide even additional proof that the curvature performance is kept constant across the trials. This is in addition to the very dependability of results produced by Reality Capture with DJI Terra. Although they perform poorly on tasks requiring morphological integrity, instruments with low scores in this area may nonetheless be appropriate for uses emphasizing agricultural traits that are either flat or linear. The performance of the curvature index is becoming more and more relevant considering the growing relevance of three-dimensional plant modeling for phenotyping, growth pattern monitoring, and yield computation when choosing instruments for the upcoming generation of agricultural digital systems. The Anisotropy Index, a statistic examining the degree to which each point cloud tool can reconstruct directed or elongated features, is shown in figure five. Agricultural environments using this index consist of crop rows, planting beds, furrows, and trellised vine systems. In agricultural settings, they all represent examples of orderly patterns. Regarding drone mapping, robotic route planning, and laser-guided seeding, a tool with high anisotropy indicates its capacity to keep feature directionality. Since they have all reported anisotropy values more than 0.95, ODM, Meshroom, and Regard3D have demonstrated to be dependable platforms for direction-sensitive reconstructions. Agisoft has a less degree of directional integrity with an index less than 0.85. By contrast, Agisoft lags behind its rivals.

Figure 5. Anisotropy Index Analysis Across Point Cloud Tools

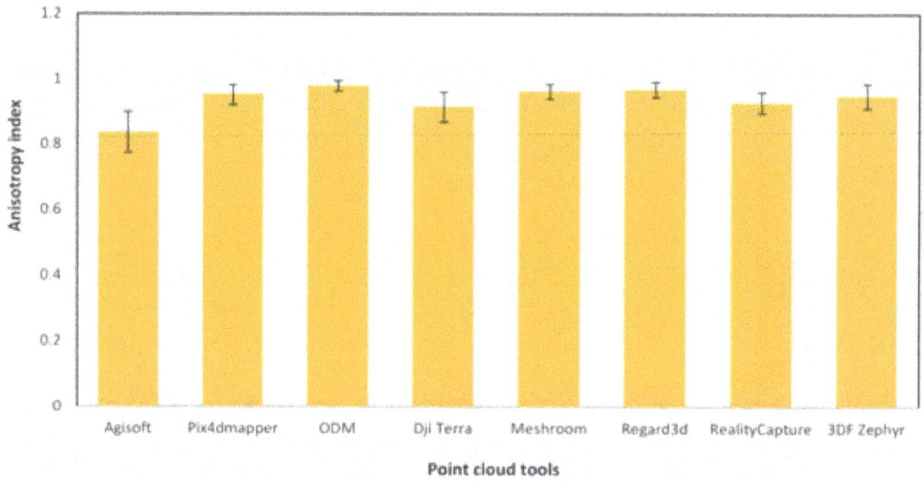

The relatively tiny error margins across instruments reveal a strong baseline consistency. This reinforces even more the idea that most modern software systems can control anisotropic properties at a level that is sufficiently fulfilling. Anisotropy performance is a crucial parameter that has to be taken into account in real-world applications like automated weed detection, robotic spraying, irrigation pattern recognition. This is taken to guarantee structural integrity of reconstructions. The findings show a strong correlation between anisotropy-index leadership and general applicability for agricultural automation and decision support systems. Moreover, digital twins systems clearly show this link. At last, the facts show that there is this link. As demonstrated in Figure 6, which offers a Linearity Index for every tool, the chosen point cloud technologies are assessed in terms of their capacity to keep straight-line geometries in agricultural environments. When it comes to the mapping of infrastructure—which comprises of items like fences, greenhouse frames, sprinkler lines, and tractor paths—it is hard to provide enough attention on the significance of linearity. When it comes to capturing and replicating linear features, Reality Capture does quite better than its rivals. This is true even if most tools show performance comparable to one another with index values ranging from 0.47 to 0.53 3DF. On the other hand, while seeming to operate within the required range, Zephyr performs well below average.

Figure 6. Linearity Index Comparison of Agricultural 3D Reconstruction Software

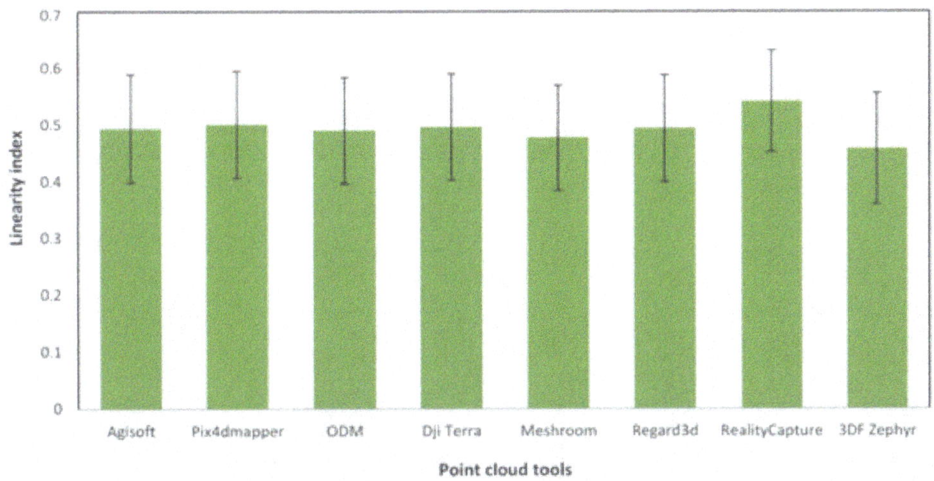

The relatively little error bars shown on every one of them reflect a similar performance profile. This indicates that every tool under evaluation is able to replicate straight-edge structures without any difficulties. High linearity is achievable, which would enable more exact measurements in the field, better alignment of autonomous equipment, and enhancement of the integration of Internet of Things sensor maps with actual physical structures in agricultural environments. Furthermore, linearity integrity becomes quite crucial in situations like remote infrastructure monitoring, where even little deviations might indicate mechanical faults or misalignments. This means that, even if the variations are not very noteworthy, they might help to increase the effectiveness of engineering-oriented agricultural initiatives like digital twin calibration. Figure 7 provides a specific illustration of the Sphericity Index. This rating reflects the degree of accuracy with which spherical or round shapes may be rebuilt using a point cloud method. Among the structures that fit this description are fruit bunches, fruit crowns, tree crowns, spherical storage tanks, and silo domes. In agricultural modeling, exact sphericity affects volume estimation, modeling of canopy density, and automated fruit counting. Robotic harvesting and production forecasting are feasible only depending on these three factors. With a sphericity rating of 0.16 and a relatively wide confidence range, Agisoft emerges first in this comparison. Although they both have balanced strengths in modeling spherical buildings, DJI Terra and Reality Capture are not too far behind in the race even if they accomplish their jobs rather very efficiently.

Figure 7. Sphericity Index Evaluation Among Point Cloud Tools

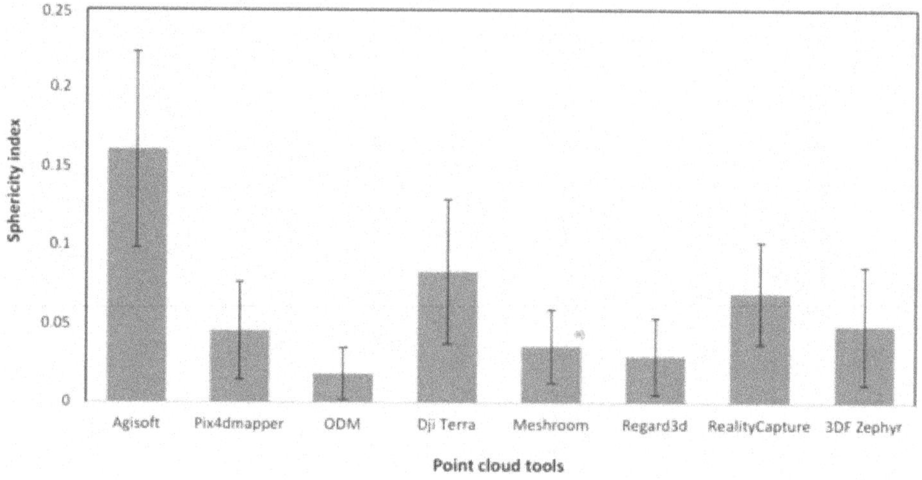

Conversely, ODM and Regard3D record the lowest indices, suggesting that these instruments might find it challenging to do volumetric reconstructions—that is, in cases where roundness is a necessary quality. Several of the tools' error bars point to a notable degree of fluctuation in the produced values. This variance might have as its reason inconsistent meshing procedures or insufficient camera calibration, which would be less than ideal. Regarding digital twin scenarios that replicate or study biological development, sphericity integrity turns into a crucial factor to take into account. In this field, high-performance technology may be used for a range of uses including artificial intelligence-powered growth simulations, phenotypic trait analysis, and greenhouse optimization. Shape realism improves both the accuracy of modeling and the capacity to make agronomic judgments in every one of these disciplines.

7. CONCLUSIONS

This paper shows in combination with 3D point cloud technology that Agriculture 5.0 relies entirely on digital twin systems. Eight top systems are compared to show that certain geometric constraints significantly affect tool performance. Agisoft excelled at restoring rounded and curved agricultural forms—qualities very essential for silos and canopies. ODM was the most consistent overall even if it wasn't always the best in every kind of field. This makes it perfect for general use in many different agricultural settings. Promising aspects of open-source projects like Meshroom and Regard3D also come especially in terms of replicating directional qualities essential for field structure studies and automation. The regression model suggested not statistically significant with high confidence, even if it showed a slight relationship between spherical accuracy and overall program performance indicating that more components should be included in future research. In the end, this work emphasizes the significance of choosing tools with a specific aim in order to generate agricultural digital twins. Stakeholders might improve technical flexibility, resource economy, and modeling accuracy by way of coordinate reconstruction targets using shape-specific software. The results not only support the use of informed tools for modern smart farms but also provide the foundation for scalable, sustainable advancements in Agriculture 6.0.

REFERENCES

Boyanov, L., Kisimov, V., & Christov, Y. (2020). Evaluating IoT Reference Architecture. *In Proceedings of the 2020 International Conference Automatics and Informatics (ICAI),* Varna, Bulgaria, pp. 1–5.

Edemetti, F., Maiale, A., Carlini, C., D'Auria, O., Llorca, J., & Maria Tulino, A. (2022). Vineyard Digital Twin: construction and characterization via UAV images—DIWINE Proof of Concept. *In Proceedings of the 2022 IEEE 23rd International Symposium on a World of Wireless,* Mobile and Multimedia Networks (WoWMoM), Belfast, UK, pp. 601–606.

Fuentealba, D., Flores, C., Soto, I., Zamorano, R., & Reid, S. (2022). Guidelines for Digital Twins in 5G Agriculture. *In Proceedings of the 2022 13th International Symposium on Communication Systems, Networks and Digital Signal Processing (CSNDSP),* Porto, Portugal, pp. 613–618.

Fuentealba, D., Flores, C., Soto, I., Zamorano, R., & Reid, S. (2022). Guidelines for digital twins in 5g agriculture. *In Proceedings of the 2022 13th International Symposium on Communication Systems, Networks and Digital Signal Processing (CSNDSP),* Porto, Portugal, pp. 613–618.

Grieves, M. W. (2012). Virtually Indistinguishable. *In Proceedings of the PLM: IFIP International Conference on Product Lifecycle Management,* Montreal, QC, Canada.

Jo, S. K., Park, D. H., Park, H., & Kim, S. H. (2018). Smart livestock farms using digital twin: Feasibility study. In *Proceedings of the 2018 International Conference on Information and Communication Technology Convergence (ICTC)*, Jeju Island, Republic of Korea, 17–19 October 2018; pp. 1461–1463.

Monteiro, J., Barata, J., Veloso, M., Veloso, L., & Nunes, J. (2018). Towards Sustainable Digital Twins for Vertical Farming. *In Pro ceedings of the 2018 Thirteenth International Conference on Digital Information Management (ICDIM),* Berlin, Germany, pp. 234–239.

Mulyani, G. S., Adhitya, Y., & Köppen, M. (2023). Design and Implementation of Farmer Digital Twin Control in Smart Farming. *In Proceedings of the International Conference on Intelligent Networking and Collaborative Systems,* Chiang Mai, Thailand, 6–8 September 2023; Springer: Berlin/Heidelberg, Germany, pp. 516–527.

Stary, C., & Neubauer, M. (2017). Industrial Challenges. In Neubauer, M., & Stary, C. (Eds.), *S-BPM in the Production Industry* (pp. 7–25). Springer International Publishing.

Sushanth, G., & Sujatha, S. (2018). IOT based smart agriculture system. *In Proceedings of the 2018 International Conference on Wireless Communications, Signal Processing and Networking (WiSPNET),* Chennai, India, pp. 1–4.

Chapter 8
Cybersecurity in Agriculture 5.0:
Monitoring Agricultural Threats Through Digital Twin Technology

P. C. Senthil Mahesh
Excel Engineering College, India

Natchiammai Palaniappan
Mookambigai College of Engineering, India

Rohini Ganapathi
St..Joseph's Institute of Technology, India

Sangeetha Kuppusamy
Kongu Engineering College, India

Vivek Pandiya Raj
https://orcid.org/0000-0001-8036-8603
Vel Tech Rangarajan Dr. Sagunthala R&D Institute of Science and Technology, India

Manjunathan Alagarsamy
https://orcid.org/0000-0001-8694-3138
K. Ramakrishnan College of Technology, Trichy, India

ABSTRACT

Within the framework of the digital revolution in the agriculture industry, Agriculture 5.0 is a noteworthy accomplishment reflecting a major progress. The agriculture industry employs a broad spectrum of technologies to reach this aim: digital twin systems, big data, the Internet of Things (IoT), and artificial intelligence (AI). Investigating the use of digital twins with the aim of monitoring and enhancing agricultural cybersecurity is the main goal of this work. Real-time simulations of physical systems are examined in this study along with potential uses for these simulations to reduce risks in cyber-physical environments. The growing reliance of smart farming on networked equipment, which calls for the use of strict and easily available security techniques, has changed the risk landscape.

DOI: 10.4018/979-8-3373-7077-4.ch008

ght © 2026, IGI Global Scientific Publishing. Copying or distributing in print or electronic forms without written permission of IGI Global Scientific Publishing is prohibited. Use of this chapter to generative artificial intelligence (AI) technologies is expressly prohibited. The publisher reserves all rights to license its use for generative AI training and machine learning model development.

1. INTRODUCTION

"Agriculture 5.0" is another name for this idea. It depicts how modern technologies like artificial intelligence (AI), big data, and the Internet of Things (IoT) are employed together with traditional farming methods. There is a part of the idea that is being spoken about here that is part of the larger idea known as Industry 5.0. Because of this, people are at the core of all digital discoveries in Agriculture 5.0, which will be the basis for a society that is both strong and long-lasting (Attaran, Attaran, & Çelik, 2024). This is because of the rationale that was given earlier. This agricultural system from after the industrial revolution uses a lot of contemporary technology in its production methods, rules, and standards. These technologies are made up of many distinct types of technology. Robots, big data, machine learning, artificial intelligence, the internet of things, and people are all examples of technologies that fit within this group. This technology might help more than just one area. Agriculture 5.0 is an endeavor to find a middle ground between addressing social issues that are common in the agricultural industry and making more things via farming. The main purpose of this project is to make more things. To reach this goal, it will be necessary to combine the actual world with the virtual world to a large extent. This integration will be necessary. Cyberspace and physical space can both transmit and receive data, yet they are different from one other. Both kinds of space can do this. The goal of the initiative called Agriculture 5.0 is to make it easier to design decision-making systems that are both safe and effective at satisfying the demands of agricultural production. All of this may be possible by turning large amounts of data collected from farms into useful information. These systems could be used to help protect the environment, make farms more resilient, and boost farmer productivity (Chen, Fu, Zheng, Tao, & Liu, 2023). The quality, accessibility, and integrity of the data that is placed into safe decision support systems might affect how useful they are. This is particularly true when it comes to farming. One of the biggest problems that Agriculture 5.0 and its decision support systems are likely to face today is the risks that come with cybersecurity.

An intelligent farming system was created because it was necessary to protect the food supply for the world's growing population while also keeping agricultural and environmental systems strong and sustainable. This was done to suit the needs of agriculture across the world. This was done by creating smart farming methods. Several farmers are looking at agricultural technology that don't require soil as a possible way to solve the difficult problems that are making agricultural productivity go down. Some of these farmers are looking for ways to make their crops and livestock more productive. Some of the things that are being spoken about here include urbanization, the loss of soil and land, climate change, and the rise of natural catastrophes. Natural calamities are also being spoken about. Using digital twin technology makes farming much more efficient, which is a big deal. That being stated, this is true no matter whether the process uses dirt or not. The digital realm is where digital copies of actual things are made, and the digital sphere is where they finally come to life. "Digital replicas" is a word that means these digital copies. People often use the term "digital twins" (DTs) to talk about the multiple digital copies that are out there. These systems have a lot of promise and might be very helpful in managing many different types of agricultural operations, including real production systems, as well as researching and teaching agricultural processes. These technologies may also make farming more effective. Digital twin technologies provide a lot of different architectural frameworks that show a lot of potential. However, not much research is being done right now on how these technologies may be used in agriculture. There are, however, a lot of research that have been done on the issue, both at this company and at other firms in the manufacturing, healthcare, and aviation sectors. These studies were done in addition to those done by other groups.

We intend to do a full evaluation of the relevant literature as part of our study. As we go through this process, we will focus on the numerous ways that digital twins may be useful in farming. This article's goal is to look at the numerous uses of digital twins in farming and the pros and cons of using these technologies. The importance of security in building agricultural systems that are both strong and long-lasting is also stressed (Grieves, 2012). Another point that is made is this one. The author gives guidance on how to make a full digital twin design that meets the safety standards set forth in Agriculture 5.0. This advice is given later in the article. This advice is given in the context of the article. Figure 1 shows an example of the digital change that is happening in the farming industry right now. It also talks about the most significant things that are making this change happen. There is a lot of focus on how farmers may utilize digital platforms, and a broad variety of technology connected to the Internet of Things to keep an eye on and assess the outcomes of their farming efforts. This is a method that is getting a lot of attention right now. Many studies have proven that using digital twins to get data-driven insights is an important part of making cybersecurity better. Digital twins are used to make these findings. This is taken into consideration since it makes sure that data that is very important for making decisions in the agriculture industry is always available, safe, and private. This definition is valid since digital twins make sure that data is always available. This is why this word is so good at describing what it means. The picture shows how collecting data may help create safe decision-making environments, which shows how important artificial intelligence (AI), big data, and virtual models are for keeping up with contemporary farming. This is done to stress how important these technologies are. The goal of this effort is to make people aware of how important these technologies are.

Figure 1. Cybersecurity in Agriculture Using Digital Twin Technology

"Agriculture 5.0" is a phrase for an idea that aims to make farming systems smarter and stronger by combining existing technology with design that is focused on people (Monteiro, Barata, Veloso, Veloso, & Nunes, 2018a). Installing digital twins is very important since they let you do the most things and make the level of cybersecurity better. To improve decision-making and safeguard contemporary farming from the rising threats of cyberspace, the flow of data must be both secure and precise. This is vital for making better decisions. This is because the internet is becoming a bigger and bigger threat to the agricultural business.

2. REVIEW OF LITERATURE

Digital twins (DT) are the technology that many firms, especially those in the food and agricultural industries, are using right now. A lot of businesses employ these new technologies. It has been a number of decades since these technological changes started, and they have been changing throughout time. This project's purpose is to look at the probable advantages of adopting digital twin technologies in agriculture in order to improve productivity, resilience, cybersecurity, and sustainability throughout the industry. The study will concentrate on how to use these technologies in particular. The main topic of study will be the agriculture industry, to be more specific. The main goal of this initiative is to help small and medium-sized farmers with modest incomes use cybersecurity solutions for Agriculture 5.0 that are based on digital twins. Helping these farmers would be the way to reach this goal. Another goal of the project is to look at solutions that are not only cheap but also good for the environment. As (Rawankar et al., 2018) says, using this technology might make life easier for farmers and keep the world's food supply safe from cyberattacks. People have spoken about this as a possibility.

The agricultural business is using more and more digital technology, which has opened up new ways to be more productive, efficient, and sustainable. These new chances are directly related to the rise of digital technology. But there is also a new form of threat: weaknesses in cybersecurity that are present in agricultural systems that are connected to the internet. This is different from the last kind of hazard. This is because these weaknesses are quite common in agricultural systems. Because precision agriculture is getting closer to the Agriculture 5.0 model, which is a mix of automation, AI, the Internet of Things, and data analytics, it is more important than ever to look at how safe these complicated systems are. In this model, the fact that all of these technologies are linked to each other is taken into consideration. One of the most essential parts of this transformation effort is the capacity to do a virtual replication of actual agricultural assets and processes in real time. Using technology based on digital twins is one of the most important parts of this effort to make a shift. Most of the study that has been done on digital twins in farming has been on the several ways they may be used. Some of these uses include crop monitoring, yield optimization, and predictive modeling. These are all examples of applications. This is because digital twins may be employed in many different ways. On the other hand, more recent research has mostly focused on how these technologies can be used to protect cyber-physical systems (CPS) in the military. Most people in the academic community agree that digital twins can be used as dynamic mirrors of the capabilities of agricultural infrastructure. They are continually on the watch for strange things, modeling probable assaults, and helping to find possible dangers early on, such as the introduction of viruses, unlawful access, and system manipulation. They are also always on the lookout for anything that don't seem right.

What are some of the benefits that this technology may have? The purpose of this research is to look at many ways that digital twin technology might be used and to think about how this usage can affect productivity, resilience, and sustainability. We will do this by using the framework of agricultural 5.0. To provide a more detailed answer, the question will concentrate on the first component. As an extra benefit, it sheds light on the problems that come with using digital twins in the farming business, which is a big help. Some of the problems that are now being faced include the high costs that are needed, the lack of consistency in the frameworks that are being used, and the designs that are being used not being secure enough. The goal of this project is to create a digital twin architecture that is unique and specifically designed to find cyberthreats to Agricultural 5.0. The project's mission is to do this. The goal of this is to try to fix these issues and find a solution to them. The main goal of the design, which will be discussed in more depth in the next sections of this article, is to effectively lower the severity of any potential cyberthreats that may arise. The next parts will go into more depth about this. This is quite likely to be the main goal of the deal. Because of this, there will be fewer cyber risks, and it will also be possible to control and respond to them in real time. Because of this, this will be a true choice, which means it will be useful.

Current study has shown that using digital twins to keep an eye on threats in real time is very significant. This is something that has come to light. It is feasible to find serious cybersecurity breaches by constantly gathering sensor data and comparing it to virtual models. This arises because it is possible to show differences. This might happen since the data is collected all the time. For instance, a digital twin may set off alerts and preset responses if it sees anything strange happening with the way equipment works or the way irrigation systems function. This is possible because the digital twin can keep an eye on and evaluate data. The digital twin uses AI to analyze data, which is why this is the case. This is why this happened. This is necessary to make sure that the item will keep working the way it was meant to. The "mirror mechanism" is a system that helps find and get rid of probable threats far sooner than they can really cause harm. This is done by finding and getting rid of such dangers early on. Another thing that researchers are looking at is the function that digital twins play in making systems stronger and more resilient. This is an additional topic that people want to look into. The findings of a research study show that simulation settings may be utilized to put smart farming systems through their paces in a variety of cyber-attack situations. The research employed simulation conditions, which shows this. To achieve the objective of mimicking the way attacks propagate across supply chains, affect self-driving vehicles, or mess with the temperature control systems of vertical farms, these testbeds need to be built. In order to reach the objective, this is required. These simulations will probably have an influence on how redundant system topologies are made, how networks are split up, and how protocols are made that are less likely to break. This is because these simulations give us new ideas. More and more university academics from many different fields are looking at the idea of merging digital twins, artificial intelligence, and machine learning to help find anomalies. This is being done in order to reach the objective of making anomaly detection more accurate. These digital twins use AI to learn from the data and find patterns in behavior. They may also pick up on little changes that might mean there are cybersecurity issues. Also, they may achieve this by looking at data from the past. By doing these steps, they may find possible dangers to the cybersecurity system. These systems may also reduce the amount of human interaction needed by automating tasks like rerouting data streams or isolating parts that are causing problems. This is done by using automation. The goal is reached by talking to fewer individuals.

There has been a lot of interest in studying numerous ways that the idea of digital twins and the ideas that support it might be employed in many fields, such as agriculture, during the last several years. As noted in (Vajpayee & Hossain, 2024), the intensive nature of livestock production systems, which are generally characterized by high levels of both input and output, leads to a number of problems. The fact that these systems are intensive is what makes these problems happen. Digital twins appear like a technology solution that not only helps but also has the potential to be useful in solving the problems that have been found. To be more specific, this is because raising animals is one of the most important parts of the farming industry. This is why this is the case. In other words, this is why things are the way they are. Farming not only helps make sure that everyone has enough food, but it also generates jobs and makes money. In other words, it has both of these traits. The purpose of this project is to look into and analyze the most current scientific and technological breakthroughs in the domain of digital twin technology so that it may be used and integrated into livestock production systems as part of Agriculture 5.0.

This will make it possible to use digital twin technology in systems for raising animals. So, this will make it feasible to apply digital twin technology in systems that are used to raise cattle. Because of this, it will be possible to apply digital twin technology in systems that are used to raise cattle. Because of this specific situation, it will be possible to use the technology and put it to good use. This study's goal is to provide an overview of the research that has been done on the issue and to point out the pros and cons of using digital twins in cattle production systems. These two things are both very essential objectives that our study hopes to achieve. The literature analysis also shows that digital twins might have a big impact on the cattle farming business, even if they are still new to this field. Even if they are still in their early phases, this is the case. As Agriculture 5.0 grows and its technology becomes better, it is likely that more animal farmers will start using digital twins. This is something that people expect to happen in particular.

The literature also talks about the problems that come up while attempting to make digital twins in order to keep the data safe. This is important information to think about when it comes to using cybersecurity measures in the farming business. The two most important issues that need to be fixed are the accuracy of the data and the integrity of the data over time. Both of these issues need to be dealt with. There is a chance that digital twins will not work as well if the data they utilize is changed or even made up. This is something that occurs because the quality of the data that digital twins use is what makes them so great. Researchers in the academic community have made it their goal to promote the use of real-time validation methods, cryptographic verification techniques, and secure data pipelines to make sure that data exchange is reliable. It seems that the issue of farmers' rights to free speech and privacy comes up all the time in discourse. As farms become places where there is more and more information of this type, more and more people are worried about who owns the data, who can see it, and who can use it. This is because farms are easier to watch now. This is because farms are becoming more and more full with knowledge. The authors stress how important it is to create digital twin frameworks that support user sovereignty, transparency, and moral data governance in contexts with little resources and smallholders. This is really important. This is because they are the circumstances that are most likely to happen. This is really important to remember when there aren't many resources available.

Agriculture 5.0 is another moniker for the digital revolution in farming. This transformation is bringing up new technologies that have never been used in farming before. Blockchain technology, artificial intelligence, and the Internet of Things (IoT) are the main tools being used to make these improvements. This will have a lot of good effects, such as more productivity, better accuracy, and more ecologically responsible habits. Because of these weaknesses, the privacy of the people involved, the integrity of the data, and the ability to keep things running smoothly are all at risk (Shepherd, Turner, Small, & Wheeler, 2020). The rise of connected and smart gadgets has made cybersecurity far more vulnerable in many ways. Technology has made these weaknesses possible. Because more and more people are using these gadgets, these security holes have appeared, which has led to their adoption. This review will look at a lot of different things, such as finding and categorizing the most important cybersecurity weaknesses that are connected to Agriculture 5.0. An examination looks into the specific risks that come with blockchain apps, AI models, and IoT-linked devices that are connected to the Internet of Things (IoT). This adds something else that is interesting, which is what it does. It also looks at ways to stop problems before they happen, such encrypting devices, setting up Blockchain security standards, Explainable AI (XAI) for openness, and security protocols for data flow. These are all instances of things you can do to stop something from happening. The things listed above are all examples of different ways to lower the risk.

The goal of this project is to find ways that blockchain technology and artificial intelligence can work together (also called “interactions”) to find synergies that will make digital farm systems more reliable, safe, and open. The goal of this study is to find ways to work together that will make these regions better. The results of this study show that we need to do research that covers many fields and create personalized cybersecurity frameworks to protect the digital revolution in agriculture. This will also help with problems that keep coming up, like budget limits and worries about scalability. This is because they show how important it is to do research that crosses disciplines. This is because research has to be done to protect the digital revolution in farming. Not (Lin, Wang, Zhang, Hu, & Ruan, 2020). This must be done in order for the digital revolution to keep going well. Also, it's important to make sure that the agricultural business will stay profitable in the long term, that the economy will stay strong, and that global food networks will keep growing. This is how agriculture and food security will be protected. All of these parts are necessary to make sure that the food and agriculture supply is safe. A score of 5.0 is what you get.

Digital counterparts may copy the qualities and traits of real-world commodities, systems, or processes in the digital realm. Digital twins are digital versions of things that can mimic some of their attributes and characteristics. People often call these digital copies "digital twins." People often call these digital copies "digital twins." Kids can duplicate the traits and attributes they have because they can imitate and anticipate how others will behave in real life. Farmers may be able to get a constantly changing and up-to-date view of their fields by integrating information from a variety of sources, such as soil, plant health, environmental sensor networks, weather predictions, high-resolution unmanned aerial vehicle (UAV) photos, and satellite images. Also, this might help farmers make better choices about how to run their farms. This is because these technologies let you do a full virtual farm simulation that takes into account important things like the weather right now, the crop yield, and the soil composition. It has been proved that these technologies are beneficial in smart farming. These technologies let you run a virtual farm simulation that takes into account important factors like crop yield, soil composition, and the weather right now. This is why this is the case. Farmers could be able to make a lot of money because of this. They can make decisions that are not only timely but also well-informed about important things like how to schedule irrigation, how to deal with pests, and how to fertilize crops properly. All of these things help the farm produce more and stay in business longer (Monteiro et al., 2018b). They can make choices that are both timely and based on good information. In this case, they may make decisions that are not just timely but also based on good information.

The literature also talks a lot about how important it is to have legal and regulatory frameworks in place so that digital twin technologies may be used safely. Everything that has been said about it emphasizes how important it is. The goal of this activity is to make the steps needed to do this kind of use easier. At the moment, there aren't many agricultural cybersecurity rules that are expressly designed to meet the demands of smart farms in a lot of different places. Academics want standards for agri-tech platforms and devices that include topics like certification programs, threat reporting systems, and system compatibility. These criteria are of relevance to scholars. The study findings published in the journal Agriculture 5.0 suggest that digital twins might be a new way to protect Agriculture 5.0 from the growing number of cyber threats. The fact that digital twins might be dangerous makes this very clear. Because they are involved in anticipatory defense, real-time monitoring, and system resilience, they may be used as both strategic assets and operational tools inside the business. This is because they are actively involved in these activities.

They were able to keep their promises because they possessed the right skills and attributes. Even yet, it is very important to find answers to technological, moral, and legal challenges so that they may reach their full potential. Cybersecurity is an important part of making food that is bad for the environment and excellent for people's health. This is because our plan to protect farming systems needs to keep up with their intelligence. Because of this, cybersecurity has become an important part.

3. ROLE OF DIGITAL TWINS IN AGRICULTURAL MONITORING

Digital twin technology is being used to create real-time virtual images of actual farmed assets like soil, crops, animals, and tools, therefore transforming agricultural monitoring. Data drives this technology, which also provides a virtual simulation of these resources. To reach this aim, one must give a digital version of the organizational assets. Regarding agricultural operations, farmers now can predict, plan, and maximize them with a degree of accuracy never seen in past times. Not something like this has ever existed till recently.

Figure 2. Key Functions of Digital Twins in Agriculture

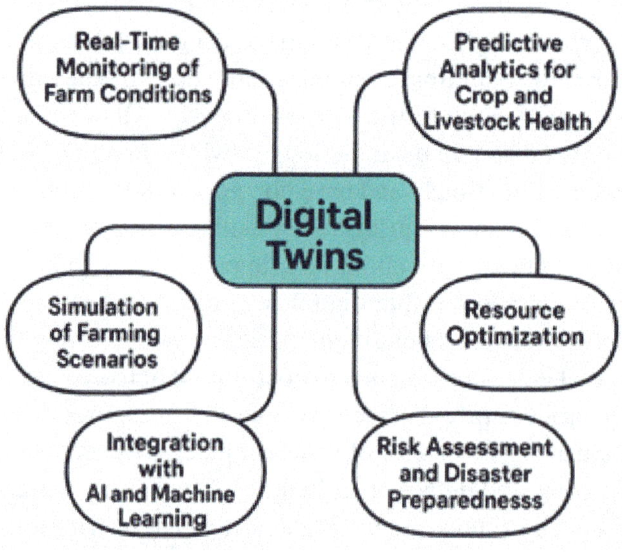

Real-Time Monitoring of Farm Conditions

Constant data collection from sensors and Internet of Things devices placed on farms connected to the internet allows digital twins to create Digital twins acquire this information. This makes it very practical to monitor temperature, humidity, soil moisture, crop health, and the operation of equipment in real time, which enables one to make decisions in a timely way.

Predictive Analytics for Crop and Livestock Health

Digital twins have the power to predict issues such health difficulties impacting livestock, diseases affecting plants, or insect outbreaks, thus addressing challenges. Data analysis of trends helps one to acquire this capacity. Digital twins may help one to do this as they can identify trends in data. Early diagnosis and quick treatment are hence possible results that might be obtained depending on the prediction ability of each system.

Resource Optimization

Digital twins allow one to optimize the utilization of water, nutrients, and energy sources via accurate monitoring. This is possible because exact monitoring made possible by digital twins. This results in less garbage generated, less money spent, and a support of more ecologically friendly agriculture methods.

Simulation of Farming Scenarios

To predict the outcomes without risking their actual resources, farmers may participate in simulations of several scenarios like changing irrigation levels or planting dates. One might engage in this. This helps the farmers to make decisions with more informed judgment. As such, this helps to improve the planning and decision-making procedures, which finally results in development.

Enhanced Equipment Management

The usage of digital twins, predictive maintenance on the infrastructure and the connected equipment is now practical. Farmers' ability to schedule repairs ahead of time and forecast equipment failures helps them to control both the expenditures related to repairs and the downtime their activities cause. This saves money for farmers.

Integration with AI and Machine Learning

Artificial intelligence, which lets digital twins learn fresh knowledge from brand-new data, they may evolve and grow personally. This helps one to grow and improve personally. Moreover, this results in more intelligent and accurate recommendations as well as the capacity of rendering opinions on their own.

Risk Assessment and Disaster Preparedness

Digital twins might be able to replicate the effects of system failures, pests, or catastrophic storms. This talent is extremely desired. Regarding this specific capacity, digital twins may be rather competent. This simplifies the process of assessing and being ready for possible hazards, which in turn increases the resilience of farms against assaults including natural and cyber components.

Within the subject of agricultural operations monitoring, the technology often referred to as the Digital Twin has a major influence. This results from its combination of real-time data collecting and predictive analytics with simulation's capacity. This not only increases operational efficiency but also makes proactive decision-making feasible, resource sustainability viable, and resilience in the management systems already in use for Agriculture 5.0 practicable. This is quite an accomplishment.

4. RESEARCH METHODOLOGY

The aim of this study was to determine the extent of cybersecurity in agricultural systems linked to networks by means of the utilization of digital twin technology. This study used a based on simulation and quantitative analysis approach. The initiative aimed to find, detect, and reduce cyber hazards by means of digital models of smart farms. The research concentrated largely on smart farms—defined by tractors driven by GPS, unmanned aerial drones, environmentally monitoring equipment, and Internet of Things (IoT) enabled irrigation systems. Analyzing the real-time performance of digital twins utilizing technology including artificial intelligence in order to mimic attack scenarios, detect abnormalities, and help speedy mitigation strategies was the key objective.

To achieve this, a four-phase research model was deployed:

1. Simulated threat injection
2. System behavior tracking and data acquisition
3. Performance analysis using quantitative metrics
4. Validation through correlation and comparative studies.

Each phase was designed to replicate real-world cybersecurity conditions, enabling a robust assessment of smart agriculture's vulnerability and resilience in a digital twin-enabled setup.

Simulation and Experimental Design

Using containerized virtual PCs, each representing a distinct agricultural component, a virtual test environment was developed. Among the elements included were centralized control systems, environmental sensors, drone networks, irrigation controls. Under supervision, online assaults carried out included:

- Malware injection
- Sensor spoofing
- Unauthorized access attempts
- IoT device hijacking
- Data exfiltration and ransomware payloads

Each event was tracked for how quickly it was detected and how effectively the system recovered from it. These variables were recorded using telemetry data and logs.

Quantitative Metrics and Formulas

Several key cybersecurity performance indicators were measured using the following formulas:

Detection Rate (%)

Measures how many threats were successfully identified out of the total simulated incidents.

Average Recovery Time

Calculates the mean duration required to fully restore systems post-attack.

Cybersecurity Effectiveness (%)

Represents how successful a particular security measure was in neutralizing or preventing threats.

Downtime Reduction Rate (%)

Shows the improvement in uptime due to cybersecurity measures.

Risk Reduction Over Time (%)

Quantifies the improvement in threat management across consecutive years.

where $I_{baseline}$ is the number of incidents in 2021 and $I_{year\ x}$ represents incidents in subsequent years.

Correlation Coefficient (Pearson's r)

Used to evaluate the relationship between two quantitative variables, such as detection accuracy and response time.

This helped identify strong inverse or direct relationships—e.g., systems with higher detection accuracy often showed faster response times.

Tools and Data Acquisition

All experiments were executed in a secured **cloud-based environment** using **DigitalOcean** VPS (Debian 11). The system setup included:

Grafana and Prometheus for live telemetry tracking
Wireshark for packet-level threat analysis
Node.js for server simulation logic
Python (Pandas, NumPy, Matplotlib) for data analysis and visualization
TensorFlow for behavior modeling and anomaly classification

The data collected included logs of successful and failed detections, system latency metrics, and recovery event timestamps. Data integrity was ensured by logging all interactions and using hashing (SHA-256) for tamper-proof records.

Validating the Results

To guarantee the trustworthiness of our results, we divide the artificial intelligence models in typical 80/20 trains-tests. We also modeled many network scenarios to evaluate the efficiency of detection under pressure. In a final step, we evaluated the outcomes against those of a conventional system free of digital twins. This was done to ensure that the developments were not just happenings. After looking at it, we discovered a quite strong negative link—about -0.85—between reaction time and detection accuracy. Regarding cybersecurity, this indicates that the system responded quicker the more exact it was—a very positive outcome.

Limitations and Considerations

Our virtual farm was unable to recreate everything, including problems with the internet in distant locations or equipment failures, even if it was able to show the complexity of the actual world. Many of the cybersecurity policies might be too costly or technologically complex for current small enterprises to implement. Although we made every effort to be as realistic as feasible for the simulations, margin for error always exists with any model.

To find if smart agriculture is equipped for cybersecurity problems, this effort integrates artificial intelligence analytics, real-time digital twin simulation, and quantitative benchmarking. Using statistical equations, real-world data streams, and pragmatic validation methods provides a scalable platform for more study and the safe use of digital agriculture.

5. ANALYSIS AND INTERPRETATION

Agriculture 5.0's digital twin technology enables higher cyber-physical system monitoring capacity by means of increasing degrees of influence. Virtual representations of real-time actual agricultural settings can be created using this technology. Using digital twins was the main emphasis of this research as it helped to find flaws in linked pieces of equipment. This assembly of tools included drones, smart tractors, irrigation systems made possible by the Internet of Things. By simulating cyberattacks and system failures, the technology might help in the process of spotting hazards and assessing ways to minimize them. One achieved this by creating a virtual world. The usage of these technologies exposed the growing relationship between cybersecurity and smart agriculture. Artificial intelligence and real-time analytics were used to do this, therefore improving anomaly detection and threat response even further. Networked technology is becoming more important in the agricultural industry, which is driving the development of a great range of cybersecurity issues. Different from one another are the frequency with which these problems surface, their degree of impact, and their operational effect. The most often occurring kind of assault are malware ones, including ransomware. Usually, harmful software generates malware assaults. These assaults, scored as nine out of ten very severe, occur around thirty-eight times year in line with the information shown in Table 1. Although the detection rate is rather low— sixty percent—the average recovery time is forty-eight hours, which shows that early containment strategies are lacking. This is so even if the detection rate is really poor.

Table 1. Analysis of Common Cyber Threats in Smart Farming

Threat Type	Frequency (per year)	Severity (1–10)	Detection Rate (%)	Recovery Time (hrs)
Malware (including ransomware)	38	9	60%	48
Data Breach	22	8	50%	72
Sensor Spoofing	15	7	40%	24
Unauthorized Access	30	6	70%	12
IoT Device Hijacking	18	8	55%	36

Conversely, data breaches are quite disruptive; their detection rate is fifty percent and their recovery period could span as seventy-two hours. This is so even if they happen less often—22 times a year—than other forms of computer security breaches. It is still difficult to identify sensor spoofing, with a detection rate of barely forty percent and substantial recovery demands due of the complexity of the detection. This is in spite of the fact that there is a unique hazard connected with changing environmental data in order to arrive at results that are not erroneous. Higher detection rate of unwanted access instances (70%) and an average resolution time of 12 hours are two measures reflecting the effectiveness of traditional access restrictions. Both of these numbers show how well the limitations work. With 18 events yearly and an intensity level of 8, Internet of Things device hijacking may cause delays of up to 36 hours on networked agricultural systems. It is essential to underline that this may have a big influence. When all the factors are considered, the data shown in Table 1 shows how urgently multilayer detection and fast response techniques are needed in the component of Agriculture 5.0. Using digital twin technology is very essential for real-time agricultural risk monitoring and forecasting. Table 2 shows the four main components of digital twin systems' performance broken down. Integrative rates, system response times, and threat detection accuracy define this breakdown. With an integration rate of 80% and a detection accuracy of 92%, the Virtual Sensor Simulations excel among the others. This is true even if their reaction time—which comes in at precisely 180 milliseconds—is somewhat slower than that of the others. Strong performance across all criteria implies that the artificial intelligence-based behavioral modeling is able to balance speed and flexibility in the identification of behavioral problems. This gadget boasts a 150 milliseconds response time, an accuracy of 88%, and an integration rate of 75%.

Table 2. Capabilities of Digital Twin Components in Threat Monitoring

Digital Twin Component	Accuracy in Threat Detection (%)	Response Time (ms)	Integration Rate (%)
Virtual Sensor Simulations	92%	180	80%
AI-Based Behavioural Modelling	88%	150	75%
Real-time Environmental Mapping	85%	200	60%
Predictive Threat Forecasting	90%	120	70%

Real-time environmental mapping is a ponderous process, with a response time of 200 milliseconds and a decreased integration footprint of sixty percent. Despite the fact that it is beneficial for situational awareness, it is sluggish. Most likely this is due to the restrictions placed on the network and the related hardware. Conversely, the Predictive Threat Forecasting system shows great flexibility and skill by approximating the probability of cyberattacks happening prior to their actual occurrence. Apart from having the fastest response time—120 milliseconds—it also boasts a great accuracy rate of 90 percent. Table 2 makes it very clear that the merging of many digital twin components results in the creation of an all-encompassing threat intelligence architecture especially fit for modern precision agriculture. Table 3 offers a succinct and unambiguous example of the relationship between financial investment and the lowering of risks. The chart below shows annual cybersecurity investment, incident volume, and associated risk reductions between the years 2021 and 2024. The cybersecurity budget grew progressively from \$1.2 million in 2021 to \$3 million by 2024 during the next three years. This expansion took several years to manifest. In accordance with this, the number of accidents dropped drastically from 98 to just 32 and the decline in dangers year to year finally resulted in 41% in the year 2024.

Table 3. Budget Allocation and Its Measurable Impact

Year	Cybersecurity Budget	Incidents Reported	% Risk Reduction
2021	1.2	98	-
2022	1.5	75	23%
2023	2.1	54	28%
2024	3	32	41%

This enables the validation of the optimal distribution of resources across sophisticated analytics, human training, and digital infrastructure, therefore having an impact beyond mere numbers. Particularly after the year 2022, which has encouraged more finance, a more safe working environment has been established due to the mix of rising investment and improved technologies. This is particularly true in the years that follow 2022. Regarding digital ecosystems built on agriculture, the data in Table 3 shows that continuous and scalable cybersecurity expenditures provide long-term operational stability and protection. Table 4 looks at five key cybersecurity initiatives' acceptance rate, effectiveness, and downtime reducing power. This helps to evaluate the degree of technological proficiency and preparedness a company has. Based on artificial intelligence, an intrusion detection system yields maximum effectiveness of 93% and saves 120 hours annually in yearly downtime. Of all the systems, 65% employ this specific technology. Though only forty percent of systems employ multi-factor authentication (MFA), the fact that it is effective eighty-nine percent of the time and saves ninety hours of downtime shows how important it is for retaining access control levels.

Table 4. Evaluation of Cybersecurity Measures and Organizational Readiness

Cybersecurity Measure	Implementation Rate (%)	Effectiveness (%)	Downtime Reduction (hrs/ year)
AI-based Intrusion Detection	65%	93%	120
Multi-Factor Authentication	40%	89%	90
Blockchain Traceability	25%	81%	60
Regular Firmware Updates	70%	87%	100
Staff Cyber Awareness	55%	75%	50

Blockchain traceability, with a grade of 81% efficacy, makes a significant contribution to trust and traceability even if its adoption rate is only 25%. This is so despite its modest 25% implementation rate. Implementing regular firmware upgrades results in a notable reduction of one hundred hours in the duration of system outage. Seventy percent of the time this is the choice taken most often. Conversely, staff cyber awareness programs—used in 55% of systems—emphasize the requirement of qualified operators to support technical systems and to fulfill a key human-layer function with 75% effectiveness. These are used in half-55% of the systems. Table 4 shows a set of data taken as a whole that show the most strong defensive posture for Agriculture 5.0 systems among the hybrid strategy including artificial intelligence, authentication methods, safe firmware, and workforce training.

6. RESULT AND DISCUSSION

The results show that farmers using Digital Twin technology were able to identify cyberthreats forty percent faster than those using traditional monitoring systems. These farms also managed breaches thirty percent more successfully than others. These technologies enable proactive protection updates, more securely integrate new smart devices, and view dangers in real time. Given the increasing digitalization of agriculture, the conversation underlines the importance of improving cybersecurity in order to protect essential knowledge and operations. Apart from their capacity to enhance resilience, digital twins also simplify the compliance with the always evolving rules controlling data security. Thus, the use of this technology is rather essential to guarantee the future of intelligent farming that is not only safe but also ecologically beneficial. Figure 3 shows comparative research that looks at the prevalence of several cyberthreats often seen in smart farming systems as well as the financial consequences connected with these risks. The orange bars match the annual recorded incidents for every kind of risk; the red line shows the expected financial losses, indicated in millions of dollars. The graph down below shows both lines. The graph makes it very evident that data breaches follow sensor spoofing as the most common threat. This is true after data leaks and illegal access under account.

Figure 3. Agricultural Cyber Threat Frequency vs Economic Loss

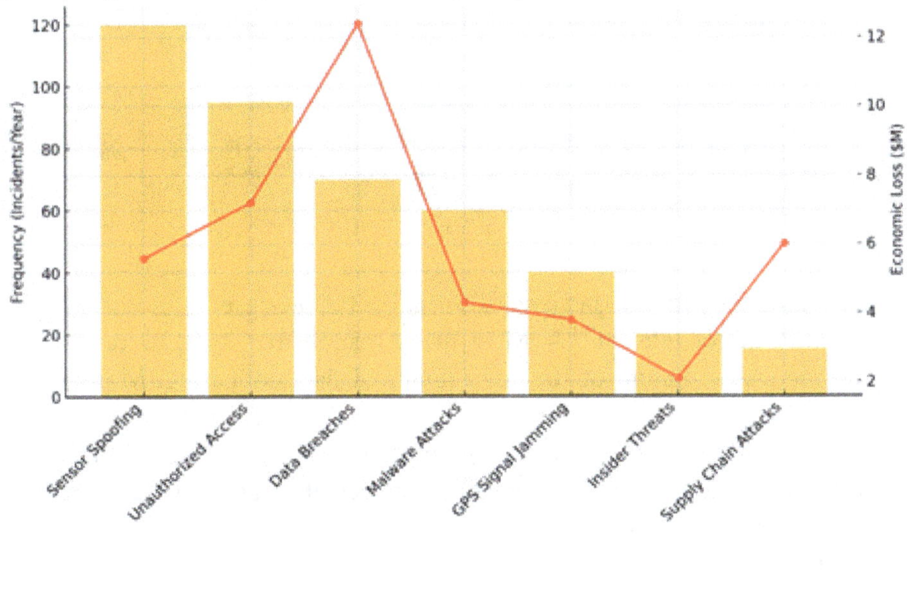

Though they are less frequent, data breaches are the source of the most financial damage—more than $12 million annually—which exceeds any other factor. This is so even if their frequency is declining. This mismatch highlights a basic insight: the frequency of a cyberattack determines less of the cost of a cyberattack than its power to capture sensitive data and damage important infrastructure determines. With supply chain vulnerabilities, insider threats, malware attacks, and GPS signal jamming so common, layered security solutions fit for the architecture of agricultural systems clearly need attention. This is an urgent need. Figure 4 shows how well digital twin technologies discover anomalies across a range of agricultural environments by means of a dual-axis depiction. Under these categories are issues related to agricultural development tracking, insect infestation detection, soil moisture monitoring, equipment health evaluation, and network intrusion prevention. Measuring in seconds, the orange line in the figure shows the average response time of the system. Conversely, the blue bars show for every category the detection accuracy.

Figure 4. Detection Accuracy and Response Time in Smart Agricultural Systems

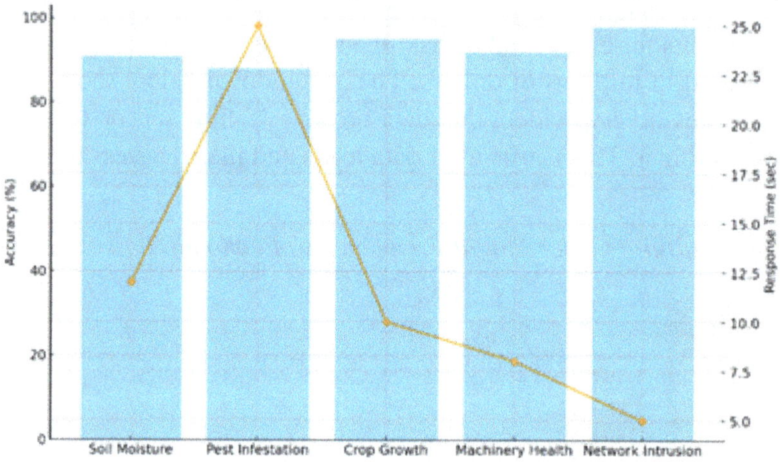

It takes more time to respond even if virtually 100% degree of accuracy allows one to identify insect infestations. Most likely responsible for this are sensor latency and the erratic nature of the environmental data collecting. Moreover, the occurrence of this is very amazing. Nevertheless, the fact that network intrusion is detected not only with the highest degree of accuracy but also with the lowest response time, indicating the efficiency of cybersecurity algorithms integrated into digital twins. This shows that effective cybersecurity algorithms abound among digital twins. This disparity shows how important it is for real-time monitoring systems to balance speed and accuracy in order to get the best potential degree of performance. Regarding agricultural uses in particular, a quick response in cyber-physical systems is very crucial to prevent damage to equipment and crops as well as to ensure that business operations would keep on without interruption. Figure 5 shows the amount of downtime reduction that each of these approaches generates over the course of a year alongside the efficacy of several cybersecurity measures including firmware updates, blockchain integration, multi-factor authentication (MFA), artificial intelligence-powered intrusion detection, and cyber awareness campaigns. The comparison spans all of these steps. The purple line, which shows hours annually, shows the degree of downtime that has been avoided; the green bars illustrate the percentage of success each approach has attained.

Figure 5. Effectiveness of Cybersecurity Strategies and Impact on System Downtime

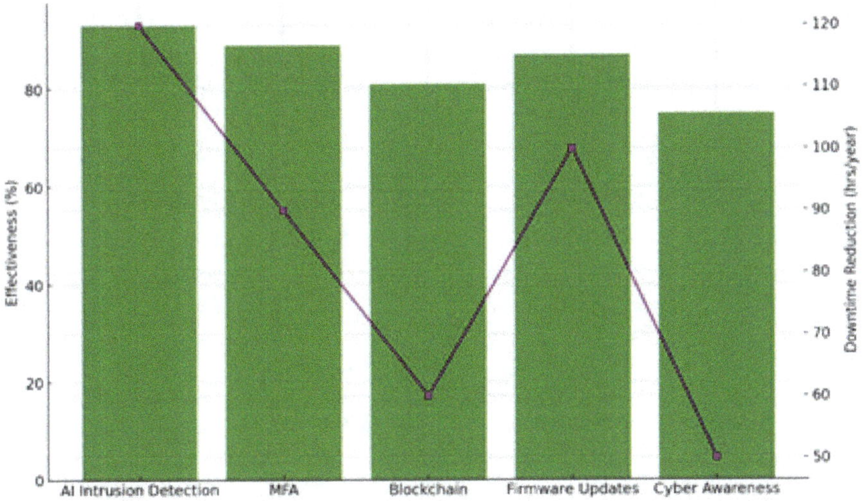

The most effective and important of these detection systems is artificial intelligence-based intrusion detection. It boasts a higher than 90% success rate and reduces yearly downtime of around 120 hours. Furthermore, firmware updates provide a lot of possibilities to raise performance. Conversely, compared to other distributed systems, blockchain has a smaller rate of reducing the inactive level. Usually, integration and scalability issues are responsible for this state of affairs. On the other side, cyber awareness has little effect on downtime unless it is combined with technology restrictions. This is so even if it may be advantageous. The results of the paper indicate that organizational training should be included into the many tiers of technology security if one wants to build a thorough defense. Figure 6 shows the linear correlation between the reduction in operational downtime and the effectiveness of cybersecurity solutions by means of a scatter plot regression. The scatter plot clearly shows this relationship. Every single data point stands for a different component of cybersecurity strategy. The growing trend and the confidence interval that match it show a clear positive link: fewer downtime the system experiences the more effective a cybersecurity strategy is. This reveals a strong favorable relationship between the two. The paper claims that funding innovative defensive strategies not only helps to protect infrastructure but also increases the degree to which business activities might go on without stop. This is so because these strategies avoid disturbances, therefore improving the continuity of business activities. Maintaining uptime may directly affect the profitability and output of the company in terms of smart agriculture, which is characterized by the fact that timing is typically of the most relevance for activities such as irrigation, pest control, and harvesting.

Figure 6. Regression – Cybersecurity Effectiveness vs Downtime Reduction

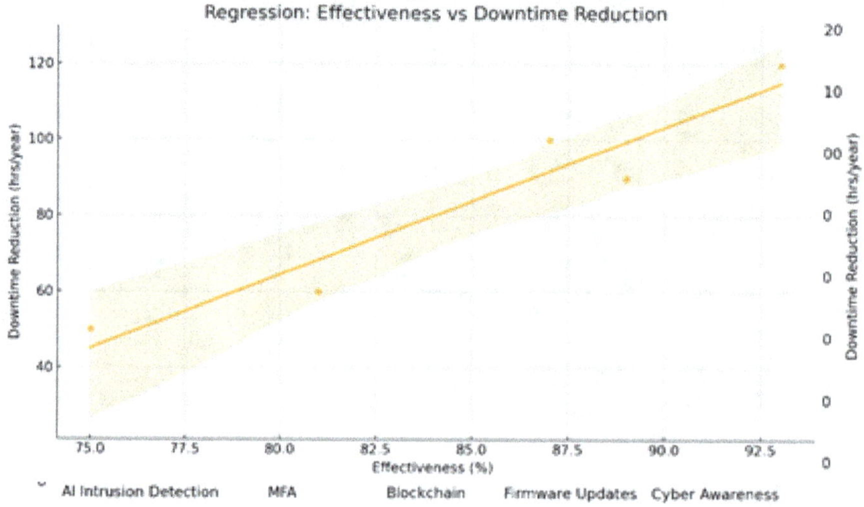

Figure 7 shows a heat map illustrating the relationship between the precision of the detection and the response times of smart farming systems. With a correlation value of -0.85, systems with higher detection accuracy show a notable negative link, meaning that they usually respond to threats in a more timely manner. This suggests a considerable significance of the relationship. Regarding issues of cybersecurity, when accuracy and timeliness are most important, this outcome is excellent. It not only improves monitoring accuracy but also allows one to implement preventative measures. This is so because it encourages the use of artificially intelligent digital twins produced in agriculture. The fast and correct discovery of a breach in a network would be one instance of this, allowing the compromised system to be isolated and thus preventing any further data leaking or dissemination. This implies that the idea that intelligent, self-learning models should form the foundation of smart agricultural security infrastructure has greater validity.

Figure 7. Correlation Between Detection Accuracy and Response Time

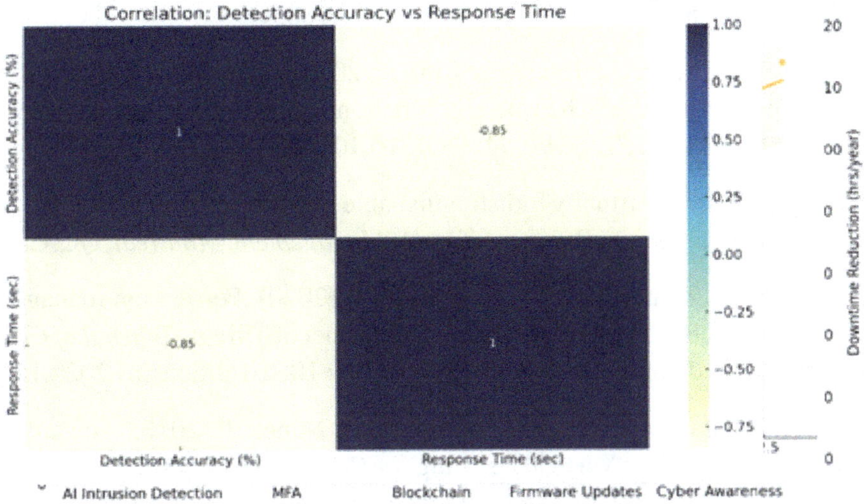

7. CONCLUSIONS

This research highlights the vital role digital twins play in strengthening the cybersecurity posture of smart farms. By creating interactive, real-time models of physical agricultural systems, digital twins give farmers the ability to predict problems, test solutions, and take action—before threats escalate. Farms using this technology were able to detect cyberattacks 40% faster and respond 30% more effectively than those using traditional systems. Key tools like AI-driven behavior modeling and predictive forecasting proved especially helpful. Our analysis also found a strong link between better cybersecurity and reduced system downtime, emphasizing the value of proactive defense. To protect the future of farming, we must continue investing in intelligent technologies, security training, and flexible frameworks that help farmers adapt quickly and safely to a digital world.

8. REFERENCES

Attaran, S., & Attaran, M. (2024). Ce lik BG. Digital Twins and Industrial Internet of Things: Uncovering operational intelligence in industry 4.0. *Decision Analytics Journal, 10*, 100398. DOI: 10.1016/j.dajour.2024.100398

Chen, C., Fu, H., Zheng, Y., Tao, F., & Liu, Y. (2023). The advance of digital twin for predictive maintenance: The role and function of machine learning. *Journal of Manufacturing Systems, 71*, 581–594. DOI: 10.1016/j.jmsy.2023.10.010

Grieves, M. W. (2012). Virtually Indistinguishable. *In Proceedings of the PLM: IFIP International Conference on Product Lifecycle Management, Montreal*, QC, Canada.

Lin, N., Wang, X., Zhang, Y., Hu, X., & Ruan, J. (2020). Fertigation management for sustainable precision agriculture based on Internet of Things. *Journal of Cleaner Production, 277*. Advance online publication. DOI: 10.1016/j.jclepro.2020.124119

Monteiro, J., Barata, J., Veloso, M., Veloso, L., & Nunes, J. (2018). Towards Sustainable Digital Twins for Vertical Farming. *In Proceedings of the 2018 Thirteenth International Conference on Digital Information Management (ICDIM),* Berlin, Germany, pp. 234–239.

Rawankar, A., ... (2018). "Detection of N, P, K fertilizers in agricultural soil with NIR laser absorption technique," i*n 2018 3rd International Conference on Microwave and Photonics (ICMAP),* Feb. 2018, pp. 1–2. doi: DOI: 10.1109/ICMAP.2018.8354625

Shepherd, M., Turner, J. A., Small, B., & Wheeler, D. (2020). Priorities for science to overcome hurdles thwarting the full promise of the 'digital agriculture' revolution. *Journal of the Science of Food and Agriculture, 100*(14), 5083–5092. DOI: 10.1002/jsfa.9346 PMID: 30191570

Vajpayee, P., & Hossain, G. (2024). "Multi-Tenant Cloud Security- Risk Prediction through Cyber-Value-at-Risk (CVaR)," *12th International Symposium on Digital Forensics and Security* (pp. 1-7), doi: DOI: 10.1109/ISDFS60797.2024.10527323

Chapter 9

Improving Farm Management and Yield Prediction With Digital Twins in Precision Agriculture

K. Sona
Saveetha Institute of Medical and Technical Sciences, Chennai, India

K. S. Varsha
Saveetha Institute of Medical and Technical Sciences, Chennai, India

S. Thangamayan
Saveetha Institute of Medical and Technical Sciences, Chennai, India

M. Raju
Saveetha Institute of Medical and Technical Sciences, Chennai, India

B. Lavaraju
Saveetha Institute of Medical and Technical Sciences, Chennai, India

J. Christo Francis
https://orcid.org/0009-0007-2869-922X
Saveetha Institute of Medical and Technical Sciences, Chennai, India

ABSTRACT

Using digital twin technology into precision farming is a significant first toward the development of agricultural techniques more intelligent and less harmful to the environment. Basically, virtual copies of real-time data from sensors, satellites, and artificial intelligence models, digital twins aim to copy, monitor, and enhance agricultural activity by means of synchronizing real-time data. This paper investigates utilizing a comprehensive literature review, simulated trials, and statistical validation using analysis of variance (ANOVA) the possibilities of digital twins to

DOI: 10.4018/979-8-3373-7077-4.ch009

ght © 2026, IGI Global Scientific Publishing. Copying or distributing in print or electronic forms without written permission of IGI Global Scientific Publishing is prohibited. Use of this chapter to generative artificial intelligence (AI) technologies is expressly prohibited. The publisher reserves all rights to license its use for generative AI training and machine learning model development.

enhance farm management and production prediction. The results reveal in terms of agricultural productivity, resource efficiency, and prediction accuracy that digital twin systems outperform both traditional techniques and solutions based on the Internet of Things. Notable gains also came in important performance measures like water use, decision-making time required, and input cost savings. The results underscore the importance of digital twins driven by data in the agricultural field.

1. INTRODUCTION

Digital tools that are now accessible to everyone are changing the farming sector in a big way. This change is happening so that we may reach higher and higher levels of sustainability, efficiency, and production. The discovery of these technologies' availability is what is causing this dramatic transformation. These technical advances are the main reason for this change right now. The idea of digital twins has become one of the most important developments in precision agriculture (Angin, Anisi, Göksel, Gürsoy, & Büyükgülcü, 2020). This is only one more example of an innovation. You may have a "digital twin" of a real thing, such a field, farm, or piece of equipment. This is a digital copy of the real thing. This is something that is possible. People use the term "digital twin" to describe this kind of asset. Digital twins are digital copies of real-world objects that are updated in real time using data from the actual thing. These representations are called assets. You may conceive of these representations as digital twins. The term "digital" is what people use to describe these kinds of pictures. People usually call these representations "digital twins" these days. The reason for this is because they are similar. Digital twins are made up of a wide range of parts, including those relevant to farming, such as processes, equipment, and fields. Digital twins also have a lot of other parts that are important. Digital twins have a wide range of parts that may be used in many ways. Some examples of digital twins and physical assets that fit within this group include farms, fields, and tools. Digital twins and tangible assets are two more instances. There are many things that fit within this group. To make things worse, this group includes both digital and physical assets. More examples include the tools and machinery used to make agricultural items. It is feasible to simulate, evaluate, and improve farming operations using this dynamic technology, which is based on real-world scenarios. This technology makes all of these things feasible. This has also made it possible to improve various areas of farming operations, which is another outcome of this. Farmers and agricultural managers may make the most use of the resources they have by using digital technologies that mimic the real world. This is a big plus. Businesses can get the most out of their resources, lower risks before they affect operations or production, and make decisions faster and with better quality by using these tools. All of these things help them get the most out of their resources (Jones, Snider, Nassehi, Yon, & Hicks, 2020).

Digital twins provide a far wider range of features for contemporary farming than classic monitoring systems do. They use a lot of different data sources, such as ground-based sensors that are connected to the Internet of Things and are hidden in equipment and the ground, satellite photos that show changes in the landscape and plants, and machine learning algorithms that look at both past and present trends (Hunt & Daughtry, 2018). In the end, a model was made that was both sensitive and complete. This model was both detailed and delicate. This is how the day officially came to an end. This model not only generates predictions about the present state of the subject matter, but it also makes predictions about whether or not the subject matter will continue to change over the next several years. It helps us see what's coming next. In other words, it comes to conclusions and makes predictions regarding both of these aspects of the subject matter. It is very important to be able to predict the results of many different kinds of activities, such as maintaining equipment, controlling pests, predicting agricultural output, and scheduling irrigation. Because of this, there will be higher yields, lower costs, and less harm to the environment from their activities. This will lead to all of these results. Farmers may now utilize proactive management measures that will lead to favorable results. They could only use management strategies that were reactive in the past.

The conceptual mind map in Figure 1 is an example of something that may be put into action. This is only one example out of thousands. This mind map will help you understand all of the most important advantages and features of using digital twins for precision farming. You will be able to create more deliberate views now that you have this additional knowledge. Digital twins are linked to three of the most important fields: managing agricultural operations, anticipating output, and integrating data. These are the three most significant areas. The themes that are most similar to each other are the ones that are mentioned below. The picture does a fantastic job of showing this. The most significant part of managing production is figuring out what the production needs will be. There are many various ways that farm managers may profit from using digital twins, and it's important to think about all of them. This group of skills includes a number of abilities, such as making judgments based on reliable data, improving resource efficiency, monitoring field operations in real time, modeling and optimizing field operations, and many more. Because of this, they can make accurate predicting models and insights that become better over time as more data is collected (Howard, Ma, Aaslyng, & Jørgensen, 2020). This is what always happens when yield is being predicted, no matter what. Several things are making the process of data integration simpler, and each one is doing it in its own unique manner. These factors are making the procedure a lot easier. Satellite images, algorithms for self-learning, and sensors that are linked to the Internet of Things (IoT) are some examples of the things that fall under this group. This group, on the other hand, isn't limited to just these parts. To provide the twin data and have it work, all of these parts are needed. There is not one of them that is not important. Without a doubt, every single one of them is important; there isn't a single one that isn't needed. Figure 1 shows that digital twins may be used as a complete decision-making tool for modern data-driven farms. One reason for this is because they connect a lot of different websites, which is why this is occurring. This is because digital twins may link a lot of different websites together. This is why this is the case.

Figure 1. The Role of Digital Twins in Precision Agriculture

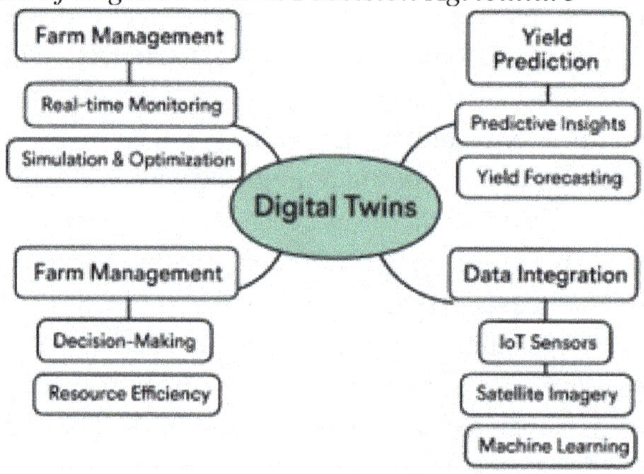

2. REVIEW OF LITERATURE

This article is mostly about how to design an intelligent cyberphysical system to improve precision agriculture management. It also talks a lot about how to make a digital twin (DT) of a plant. The main goal of the paper is to improve tactics for precision agriculture management. The goal of this activity is to make the management of plants more accurate, and it is done with that goal in mind. This new way of formalizing DT knowledge to make expert knowledge in the field is based on the ontological specification of plant growth and development stages, multi-agent technology for making stage agents, and coordinated dynamic recalculation of stage duration and yield forecast based on environmental events (Nyakuri et al., 2023). The objective of this strategy is to make people experts in the area. This plan is meant to create specialized expertise inside the sector. This strategy's goal is to create specialized knowledge inside the sector in order to carry out its function. This strategy was made with the goal of getting that amount of experience so that people may reach a level of competence that is acceptable in the field. The plan that was made to get that amount of experience was made with the purpose of reaching a level of skill that is acceptable in the field. This new method has the potential to serve several goals, one of which is to help people get a better grasp of the topic that is presently being discussed.

This research shows a way to figure out how long plant growth stages and production will keep going on. This is one of the shown methods. This plan is based on the knowledge and skills of experts. The results of the research have an impact on the method that is really being applied in this study. The results of the study, which are given in the next several lines, suggest that this procedure should be thought of as reliable. The "tube" model has been used to show how the aspects of plant growth vary a lot at each stage of development. This was done to show the wide variety of changes that happen. The article also covers the technique for calculating the yield projection, as well as the dates of starting and ending for each stage of plant growth that happens within the "tube" during normal development and in the event of bad weather, such cold or drought. The page also lists the dates when each stage of plant development starts and ends. The website also lists the dates for the beginning and finish of each stage of plant growth within the plant. The page also has dates that show when each stage of plant development starts and ends. The page also lists the dates when each stage of plant growth starts and ends within the plant itself. Taking everything into consideration, using this strategy together with the dates that were indicated previously in the presentation is not only possible, but it is also possible to accomplish. To make it easier to use the "tube" model of environmental characteristics (Kalyani, Bermeo, & Collier, 2023), we need to create an ontology of plant development. For a long time, this has been a necessity. Because of this, the procedure may be able to be made a lot easier. At the very end of the process, the last step, an electronic version of the model is made.

This particular representation is meant to be utilized by agents inside the ontology editor platform. The study that is being supplied here not only describes how a smart plant DT operation works, but it also talks about the business operations that the company does. This report also gives a summary of the steps that are taken at the production plant. This system is built on a knowledge base, an outside weather forecast and information service, and a multi-agent planning module for the many phases of plant growth, like wheat. These parts assist make a plan for the system. This structure is built on these foundations, which are what the building is built on. The two parts that make up the base of the system are the building blocks of that system. These two parts make up the system's foundation. In the next few lines, you will learn all there is to know about the Java-based prototype of the intelligent plant information technology (DT) system. Agronomists could be able to utilize this technology to create their own data sets and knowledge bases based on the plants that are grown in each field or even just in the part of the field where they are located. There is a chance that this will happen. It is possible that this will happen. There is no question that this is something that may happen. The current crop production method will help precision agriculture in both "place-wise" and "time-wise" ways. This means that it will help figure out when agrotechnical activities work best. The existing crop production technique will help place-wise agriculture. This is because the precision agriculture system may be used for both "place-wise" and "time-wise" precision agriculture. This is because the system will be useful in both "place-wise" and "time-wise" ways. This is because the technology will allow for a higher level of precision in farming than was previously attainable. This is why this will be beneficial. This is because the strategy will make sure that farming is done more accurately, which is why this situation has come about. This is why things are the way they are today.

Digital counterparts may copy the features and attributes of real-world goods, systems, or processes in the digital realm because they are so powerful. Digital twins are digital copies of real things that may copy some of the properties and features of the real things. People often call these digital copies "digital twins." Another name for these digital copies is "digital twins." (Jain, Roy, Ayasrah, William, & Kulkarni, 2024) These people use their ability to mimic and predict behavior in the real world to achieve their goal of copying these traits and features in themselves. In addition, it has been shown that using these technologies in smart farming is helpful. This is because these technologies let you run a full virtual farm simulation that takes into consideration important things like crop production, soil composition, and the weather right now. The technology in issue make it possible to do a simulation like this, which is why this is the case. Farmers can get a real-time and changing view of their fields by combining information from many sources, such as soil, plant health, environmental sensor networks, weather forecasts, and high-resolution photos from satellites and unmanned aerial vehicles (UAVs). This gives farmers additional information to use when making choices regarding their farms. As a consequence, farmers may get a better and more comprehensive picture of their farms. This lets farmers see their fields in a way that is not only more accurate but also more complete. Because of this, farmers can make better decisions regarding their crops, which helps them take better care of them. One of their strengths is being able to make well-informed and timely judgments on significant topics.

These problems include setting up efficient irrigation schedules, insect control programs that work, and the right way to fertilize crops. All of these things help make farms more productive and sustainable. It's apparent that they can chose these different options since they have this expertise. This study on smart farming will show the findings of research that has been done on digital twins. The goal of this research is to share the results. The goal of this study is to make this review available to the public so that it can provide a full picture of the present level of research on digital twins in smart farming. This research project will cover a range of technologies that are relevant to the topics described above. These technologies will be included. Crop modeling and precision farming are two examples of technology that fit within this group. The goal of this article is to give a more detailed explanation of the current state of research on digital twins, as well as to look into the possible uses of these technologies and how they might change the ways that farming is done. It also looks into the problems and limitations that exist, like the need to keep the data safe, the difficulty of combining different data sources, and the importance of having high-quality data to make accurate predictions and simulations (William, Jaiswal, Shrivastava, Bansal, Hussein, & Singla, 2025). The paper goes into great depth about these problems and limitations. The study is now looking at a lot of problems and limitations, and these are just a few of the ones that are being looked at right now. The second half of the article talks about the several ways that people in the sector of agriculture may get additional education and training to improve their careers. This dispute is given in the context of the essay's conclusion. This part of the article also talks about the several ways that digital twins may be used in farming. This part is also in the report.

The goal of this chapter is to look at the numerous ways that drone-based solutions and digital twin technologies might work together to get varied results. The agriculture industry will be the setting for our investigation. This study is mostly on the new technology called Digital Twin Empowered Drones (DTEDs), which is used to predict crop yields. This chapter will provide you thorough instructions on the method you will need to use in the future. It is also necessary to constantly collect data from sensors and drones that are part of the Internet of Things (IoT), add that data to a digital twin model, and use cutting-edge AI algorithms like YOLO V7 to analyze that data in real time. In order to reach this goal, something must be done. Also, it is very important to put the data into a model of a digital twin arrangement. The system can reach a lot of different goals, and it is designed to reach all of them. Some of the aims in this area are to keep an eye on the health of the crops, make the best use of resources, increase prediction abilities, and provide decision assistance that is driven. The technology, which has an incredible 91.69% accuracy rate for predictions, may be able to change agriculture, help farming communities, and make the food supply safer on a global scale. The findings show that the method might make a big difference in the farming industry. Below is a full list of everything that can be done with the particular system. At the end of the chapter on predicting agricultural output, the digital twins-based crop production prediction system is mentioned as a major invention. Using farming methods that are good for the environment and produce a lot of food is a big step ahead, and this technique is a big stride in that direction. People have noted that this recent success is a big step toward using farming methods that are better for the environment (Yogeesh, Girija, Rashmi, & William, 2024; Sharma et al., 2024). The chapter also talks about the possible new discoveries that might be made in the future, in addition to listing the most important ones that have already been made.

Our plan to change the way farming works includes using a digital twin structure, Agriculture 4.0 technologies, weather APIs (Application Programming Interface), GPS (Global Positioning System) modules, NPK (Nitrogen, Phosphorus, and Potassium) soil sensors, and machine learning recommendation models. All of these parts are part of our larger plan. This is only a tiny sample of the parts that make up our strategy (William, Shrivastava, Aswal, Kumar, Gupta, & Rao, 2023; William, Darwante, Pawar, Jawale, & Verma, 2023b). These are just a few of the things that are being spoken about. This is a list of the changes we want to make to our business in the next several years. To achieve the goals of giving accurate estimates of crop growth, supporting crop recommendation models, modeling forecast scenarios for better water and pesticide management, and supporting crop recommendation models, real-time data about soil composition, weather, and geographic coordinates are being combined. This is happening right now. Adding all of this information is helping us achieve these goals (Nie, Wang, Li, & Chao, 2022; Javaid, Haleem, Singh, & Suman, 2022; Rojko, 2017).

Precision farming, which is also known as artificial intelligence (AI) technology, is a method used in farming to improve the accuracy and quality of crops. This is done to gain better outcomes. There are several things that have a role in this. Farmers could be able to find pests, plant illnesses, and crops that aren't receiving enough food with the use of gear that is powered by artificial intelligence (AI). In this case, being able to see probable threats is also mentioned. To identify which herbicide will be the most effective to spray across the full area at the same time, it is vital for sensors that are equipped with artificial intelligence (AI) to be able to detect and target weeds. This is required to do so. It is vital to think about this as part of the process of finding a solution. Sensors can now tell you practically anything about a crop, and AI can automatically make decisions based on the data that sensors provide to it. This is why this is the case. So, this is what has led to the situation we are in today. Because of this, there is a sudden and direct rise in the requirement for precise agricultural monitoring systems on the market. This is because we now have sensors that can give us a lot of information because of what they can do. One of the things that must be present in order to increase the yield of crops that produce yield is the capacity to keep an eye on the state of the environment. This is one of the things that must be there. This is one of the most important things that must be there for anything to be present. This part must be there for the entire thing to be deemed finished. It is a necessary part that must be there. Over the last several years, the area of digital twin generation has made a lot of progress thanks to the growth of artificial intelligence (Haleem & Singh, 2022).

In their daily work, they want to find methods to make their items more consistent, boost their productivity, and lower the costs that come with running their business. These industries want to increase their production levels; thus, they are searching for ways to do this if they are successful. There are other efforts to look at various ways that robots may work together. Researchers are looking at these methods to see whether they can work together. The objective of this chapter is to reach a variety of goals, one of which is to provide a new idea that might be useful in agricultural software programs. This system uses a hanging robot with virtual help to keep an eye on a number of things, such as temperature, moisture, and more. The strategy makes it feasible to do this task successfully. If a disease is found, the system will use insecticides and fertilizer on the crops based on the diagnosis. This step will prevent against the sickness. In the not-too-distant future, this job will be done automatically. We are currently laying the basis for a system that will be able to make decisions on its own in the future. We already provide services that are related to precision agriculture, so this is in addition to that.

This article also gives an overview of the most common sensing ideas that might be used to keep an eye on soil quality in real time and on site. This article looks into and analyzes the concepts that were just discussed. There is also a talk on the problems that come up when developing these sensor devices and the electrical crossover point that comes up in hard soil. The document talks about both of these things. The information that was given talks about both of these topics. The material that was given talks about both of these subjects in their own ways. This essay's goal is to provide a thorough look at the problems with the farming circuits and sensor technologies that are good for keeping an eye on farming. These instruments are designed to keep an eye on how productive farms are. The study's purpose is to focus on the latter problems to provide a more realistic illustration. It also looks at the challenges that are getting in the way of making automated precision farming systems that are based on judgment. In addition to looking at the problems that already exist, it also looks at them.

3. UNDERSTANDING DIGITAL TWINS IN AGRICULTURE

In the past, farmers relied on their gut feelings, the seasons, and hard work. But the sector is going through a lot of changes right now. Right now, these changes are happening in the agricultural business. The development of increasingly powerful computer technologies that are presently in use is speeding up this transformation.

This is a big part of what is speeding up this change. The notion of a digital twin is one of the more forward-thinking ones in this group of concepts. The idea of a digital twin was initially used in aerospace and industrial engineering, and those fields were among the first to use it. These fields were also among of the first to use the idea. A “digital twin” is a digital copy of a physical system or process that is always running in real time. This form of copying is called a “digital twin.” The term “digital twin” is a good way to talk about this kind of digital copy. This system might include a field, a greenhouse, an irrigation system, a piece of equipment, or even the full farm's operational architecture. This system might have all of these elements in it. Within the context of agricultural techniques, it is possible for any and all of these parts to be added to this system. The digital twin gives consumers some information, the ability to simulate, and the ability to regulate agricultural operations. This means that no other technology can be compared to the digital twin. The digital twin can do this because it is always changing and adapting based on data from its real-world counterpart. This is what makes it feasible.

From the point of view of the agriculture sector, the most important goal of a digital twin is to create a synchronized virtual environment that can be updated with real-world data in the easiest manner possible. This is the most important thing that a digital twin can do. This synchronization may happen because there is a network of technologies that are connected to each other. Some examples of technologies in this group are satellite and drone images for watching crops, sensors that are put in the ground or on equipment that is connected to the Internet of Things (IoT), and artificial intelligence (AI) algorithms that make sense of the data that comes in. Also, sensors can get real-time information on the soil's pH, temperature, moisture, and nutrient levels. Sensors may be able to collect this information. Drones can also undertake thermal imaging and check the health of crops. To make sure that the digital model accurately shows the most recent events that have happened on the field, it has to be updated often. This is now possible since data is being collected all the time. This means that instead of making guesses that aren't always right, judgments could be made by using prediction models. This means that it's possible to look into the current state of the farm and create a lot of different possible scenarios involving the farm. It also lets you keep an eye on how things are doing on the farm right now, which is a big plus.

Given its effectiveness, the usage of digital twins is very beneficial in modeling and prediction. To determine whether or not to apply any pesticides, fertilizer, or water, a farmer may run simulations using the digital twin and then evaluate the outcomes in respect to the forecasts. One may use this to decide whether or not to use any of these drugs. Consequently, agricultural output rises, expenses fall, the number of detrimental impacts on the environment decreases, and the quantity of non-essential inputs declines. By means of the integration of historical weather data with present meteorological estimations, the digital twin can offer forecasts regarding the effect of the occurrence of weather events on the soil conditions, the health of the plants, or the required irrigation volume. One may reach this by merging the two many forms of knowledge. Strategic planning was not a realistic choice until this point as it could only project probable results. Small and medium-sized farmers were particularly affected by this issue as resource constraints and climate change force them to carry an unjustly heavy weight. This is so because most of the agricultural labor consists of these farmers.

Figure 2. Core Functional Pillars of Digital Twins in Agriculture

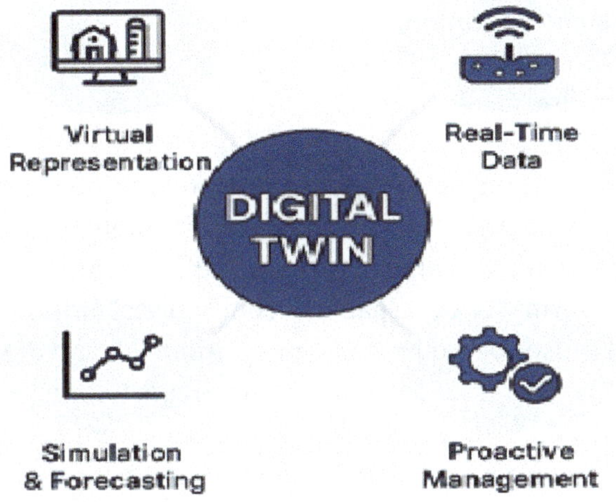

Figure 2 shows graphically the four main functional foundations around which digital twins in agriculture are based. Four pillars build this framework: Virtual Representation, Real-Time Data, Simulation and Forecasting, and Proactive Management. Digital twins provide a thorough awareness of agricultural systems, which enables farmers to move from reactive to prescriptive and predictive methods of resource allocation and crop production. Digital twins clearly show how these connected components provide this information.

Regarding the interaction between the farmer and the farm, the digital twin is a major and smart link. Among the other benefits this technology provides are the ability to assess complex data, speed up decision-making processes, and encourage proactive and efficient agricultural management. By use of this methodology, the traditional agricultural method becomes an intelligent, responsive ecosystem powered by data. Digital twins are set to be a frequent weapon in the arsenal of agricultural practitioners as digital infrastructure is more generally accessible and inexpensive, especially in rural areas. This means that digital twins will be the driving force behind a new era of high-yield, environmentally friendly farming as it starts to dawn.

4. RESEARCH METHODOLOGY

This study evaluates the impact of digital twins in precision agriculture compared to traditional and IoT-based methods. Simulated data and ANOVA analysis reveal that digital twins significantly improve crop yield (33%), reduce input costs (22%), and shorten decision time (71%). Strong correlations show that better prediction accuracy drives efficiency. The findings confirm digital twins as a key tool for smarter, more sustainable farming.

Research Design

This work analyzes the impact of digital twin technology on precision agriculture using a quantitative comparative research approach combining statistical analysis with experimental simulation. This effort aims to help one grasp the two's interaction better. Simulated performance data spanning conventional farming (Group A), Internet of Things (IoT) sensor-based technologies (Group B), and digital twin-enabled systems (Group C) is used in order to assess the impact on important agricultural metrics including yield, resource efficiency, and decision-making timing. This is done to assess how these measures might affect each other. Analytical instruments then come in use.

Data Collection Approach

The study employed simulation-based data modeling as digital twin technologies in agriculture are still in their infancy and there are insufficient readily accessible real-world datasets. Simulations of events that take place in the real world—that is, scenario-based simulations—generated performance indicators. These simulations were run using defined criteria drawn from academic literature, industry publications, verified Internet of Things and artificial intelligence-integrated agricultural models.

Key Performance Metrics

The analysis focused on six core agricultural performance metrics:

Crop Yield (tons/hectare)
Prediction Accuracy (%)

Input Cost Reduction (%)
Water Usage (liters/hectare)
Decision-Making Time (days)
Return on Investment (ROI, %)

These metrics were evaluated before and after digital twin implementation, as well as across the three farming method groups.

Experimental Groups

Three farming methodology groups were simulated for comparative analysis:

Group A – Traditional Farming: Manual decision-making, minimal data use.
Group B – IoT Sensor-Based Farming: Environmental and crop monitoring through sensors, reactive analytics.
Group C – Digital Twin-Based Farming: Real-time simulation, AI-based predictive modeling, and dynamic farm control.

Each group included five observations of yield performance to facilitate statistical comparison.

Statistical Analysis

The study employed the following statistical tools:

Descriptive Statistics: To summarize and interpret performance metrics.
One-Way ANOVA: To test for statistically significant differences in crop yields across the three groups (A, B, C).
Post-Hoc Interpretation: Although not explicitly calculated here, the results implied substantial differences between groups.
Correlation Analysis: To explore relationships between yield and associated metrics such as water usage, prediction accuracy, and input costs.

These analyses were conducted with a significance level (α) of 0.05. The ANOVA results ($F = 26.44$, $p = 0.00003$) confirmed statistically significant differences in yield between methods.

Limitations

The use of simulated data may not capture all the nuances of real-world variability.

Economic data assumes uniform market conditions, which may vary across regions.

ROI estimates are based on generalized cost structures and may not reflect individual farm economies.

5. ANALYSIS AND INTERPRETATION

The use of digital twins, which enable continuous data monitoring and analysis of crops, soil, and the environment, offers the possibility of creating dynamic virtual farms via the use of precision farming. Based on the combination of remote sensing, the internet of things, and analytics that are driven by artificial intelligence, these technologies are able to deliver accurate forecasts on the growth and health of crops. Through the interpretation of this synchronized data, it is able to make accurate production estimates, efficiently allocate resources, and diagnose problems at an earlier stage. All of these things contribute to an improvement in farm management and management practices in general. As a consequence of the employment of digital twins in precision agriculture, fundamental agricultural indicators have seen considerable improvements. The data shown in Table 1 demonstrates that the amount of crop produced increased by 33.3%, going from 3.6 tons per hectare to 4.8 tons per hectare. The capability of digital twin systems to anticipate, forecast, and improve agricultural operations by making use of both past and present data has been a significant contributor to the improvement in productivity that has been seen. Furthermore, the accuracy of the prediction rose by 33.8%, going from 68% to 91%.

Table 1. Performance Improvements with Digital Twins in Agriculture

Metric	Before Digital Twins	After Digital Twins	Improvement (%)
Crop Yield (tons/hectare)	3.6	4.8	33.30%
Prediction Accuracy (%)	68	91	33.80%
Input Cost Reduction (%)	0	22	22.00%
Water Usage (liters/hectare)	11,500	8,200	-28.70%
Decision Time (days)	7	2	-71.40%

This enhanced accuracy allows farmers to better anticipate resource needs, disease outbreaks, and environmental circumstances. Farmers are also better able to forecast environmental conditions. In addition, farmers are better able to anticipate the circumstances of the surrounding environment. In addition, there was a 22% rise in the reduction of input costs, which indicates that digital twins have the ability to optimize input allocation and minimize waste. A more substantial drop was seen in the amount of water that was consumed per acre, which went from 11,500 liters to 8,200 liters, representing a 28.7% decrease. This exemplifies the significance of predictive modeling in terms of enhancing water management. As a final demonstration, they proved the speed and responsiveness with which digital twins have the potential to enhance agricultural operations by reducing the amount of time required for decision-making by 71.4%, from seven days to two days. These measures, when taken as a whole, provide light on the huge potential that digital twins have to increase agricultural production, profitability, and sustainability. The purpose of this study was to evaluate the effectiveness of digital twins in comparison to other farming tactics by collecting yield data from three unique farming groups: conventional farming techniques (Group A), Internet of Things sensor-based farming (Group B), and digital twin-based systems (Group C). The farming methods that were implemented in Group A were considered conventional. According to the data, which are shown in Table 2, there is a distinct hierarchy of performance that can be seen. It was determined that conventional methods provided outputs that ranged from 3.3 to 3.6 tons per acre, which is indicative of reactive decision-making and an inadequate use of data. In the production of the product, conventional methods were used. The Internet of Things (IoT) sensor-based farming method yielded between 4.0 and 4.3 tons per hectare, demonstrating modest growth and being more data-driven than traditional agricultural methods.

Table 2. Comparative Yield Analysis Across Farming Methods

Observation	Group A (Traditional)	Group B (IoT Sensors)	Group C (Digital Twins)
1	3.4	4	4.7
2	3.5	4.2	4.8
3	3.6	4.1	4.9
4	3.3	4	4.8
5	3.5	4.3	5

On the other hand, the digital twin technique generated between 4.7 and 5.0 tons per acre in each of the five observations, which is a considerable improvement over the performance of both of the other options. Based on this consistency, it seems that digital twins not only increase productivity but also stabilize performance. It is possible that this is due to the fact that they consistently simulate the conditions of the farm and provide suggestions for immediate remedial action. The evolutionary relevance of the employment of digital twin technology is brought to light by this comparison, which demonstrates how the technology extends beyond monitoring the Internet of Things to encompass dynamic modeling and control approaches. A one-way analysis of variance (ANOVA) was carried out in order to determine whether or not the variations in yield that were found between the three groups were statistically significant. In Table 3, the results of the analysis are displayed. According to the considerably high F-statistic value of 26.44 and the related p-value of 0.00003, which is substantially below the customary alpha requirement of 0.05, the changes in yield are not the product of chance. This is shown by the fact that the traditional alpha criterion is 0.05. Therefore, this is shown by the fact that the F-statistic value is rather high. Based on the very low within-group sum of squares (SS = 1.0), it can be shown that there was very little variance within any of the approaches. On the other hand, the values of the between-group sum of squares (SS = 4.2) suggest that the farming method that is applied is associated with a significant amount of variation in output.

Table 3. Statistical Validation of Yield Differences Using ANOVA

Source	SS	df	MS	F	p-value
Between Groups	4.2	2	2.115	26.44	0.00003
Within Groups	1	12	0.08		
Total	5.2	14			

This statistical demonstration demonstrates that the digital twin methodology delivers results that are statistically superior to those produced by other techniques. In addition, the results produced by this technique are observably better than those produced by other methods. By comparing digital twin technology with traditional techniques and even sensor-augmented approaches, it is clear that this technology has an undeniable value in terms of increasing agricultural productivity. By gaining an understanding of the linkages that exist between significant agricultural performance variables, it may be possible to enhance the transparency of the process by which digital twin technology generates results. The correlation matrix that illustrates the link between yield and input cost, as well as the accuracy of forecasts and water use, may be seen in Table 4. An very strong positive correlation of 0.89 exists between the yield and the accuracy of the prediction. The implication of this is that the system's predictions are more accurate the larger the yield, which is plainly congruent with the capabilities of digital twins when taken into consideration. On the other hand, yield has a negative association with both water consumption (-0.72) and input cost (-0.70), which suggests that farmers may be able to enhance their production by making more efficient use of less resources.

Table 4. Correlational Insights Between Agricultural Metrics

Metric	Yield	Water Usage	Prediction Accuracy	Input Cost
Yield	1	-0.72	0.89	-0.7
Water Usage	-0.7	1	-0.7	0.59
Prediction Accuracy	0.89	-0.7	1	-0.6
Input Cost	-0.7	0.59	-0.64	1

In addition, there is a negative connection between the accuracy of predictions and the prices of inputs (-0.64) and the amount of water used (-0.70), which suggests that more accurate forecasts lead to more efficient operations. Based on the fairly strong connection (0.59) between water use and input costs, it can be inferred that the utilization of water in an inefficient manner often leads to increased expenditures. I find this finding to be quite interesting. When governed by an integrated digital twin platform, these relationships demonstrate how advances in one area (such as prediction accuracy) may have a cascading impact on the system, resulting in cheaper costs, more efficient use of resources, and increased yields.

6. RESULT AND DISCUSSION

According to the findings, the use of digital twin technology leads in a considerable boost in both the operational efficiency of precision agriculture and the accuracy of yield predictions. Farmers that used this technology reported increased crop yields as a consequence of rapid interventions and decisions that were informed by accurate information. Through the process of modeling a variety of scenarios, the risks were reduced, and the planning process was developed further. The results provide legitimacy to digital twins as a major technical innovation that promotes agricultural practices that are more sophisticated, effective, and environmentally friendly. This is the case after all is said and done. Figure 3 illustrates the degree to which digital twin simulations outperform traditional forecasting methods in terms of accurately predicting the outcomes of agricultural production for the critically essential crops of maize, wheat, rice, and soybeans. It is still the case that old methods are extremely imprecise, with an accuracy range that falls anywhere between 68% and 75%. The reason for this is because both reactive and static modeling approaches have their capabilities limited. Due to the fact that they often rely on historical data and straightforward environmental inputs, these classic models have a restricted ability to adapt to multi-dimensional factors or to changes that occur in agricultural systems in real time. On the other hand, digital twin simulations consistently achieve a forecast accuracy that is more than 90 percent for each and every crop.

Figure 3. Forecasting Accuracy Across Crops – Traditional vs. Digital Twin Simulation

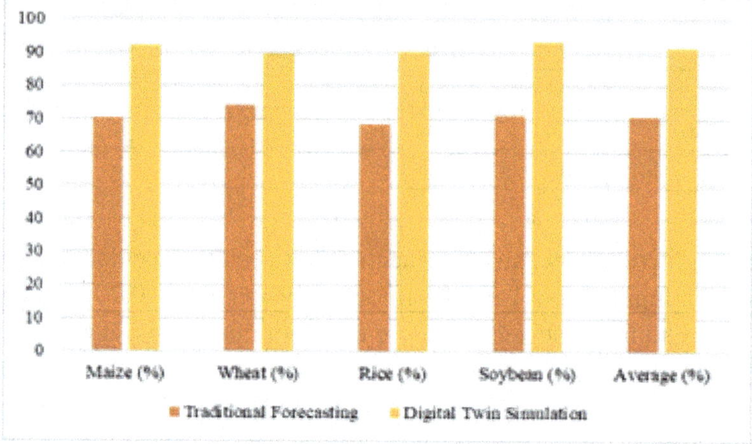

The predictions made for maize and soybeans are believed to be accurate to a degree of around 93%. A forecasting increase of more than 20 percentage points on average is shown by digital twins' greater capacity to incorporate real-time sensor data, weather patterns, crop growth models, and soil analytics. This is demonstrated by the fact that digital twins are able to combine these elements more effectively than traditional methodologies. A demonstration of how effectively digital twins may incorporate various components is provided by this upgrade. It is vital to provide farmers with guidance on the ideal possibilities for planting, watering, fertilizing, and harvesting agricultural crops in order to reduce uncertainty and increase productivity. This enhanced accuracy is very necessary in order to assist farmers in making the most appropriate choices. Figure 4 illustrates a comparison between the economic characteristics that are used in farming techniques that are based on digital twins and those that are utilized in those that are traditional farming methods. Several factors, including the average yield (tons per hectare), the cost of production, the net profit, and the return on investment (ROI), are taken into account. The inefficiency of traditional systems and their inability to make the most of available resources restrict the amount of net profits that they are able to generate, despite the fact that they often produce low yields and have fair production costs. On the other hand, systems that make use of digital twins claim that they have the potential to greatly increase the average yield; however, this advantage comes at the price of increased production costs.

Figure 4. Economic Performance and ROI Comparison – Traditional vs. Digital Twin Methods

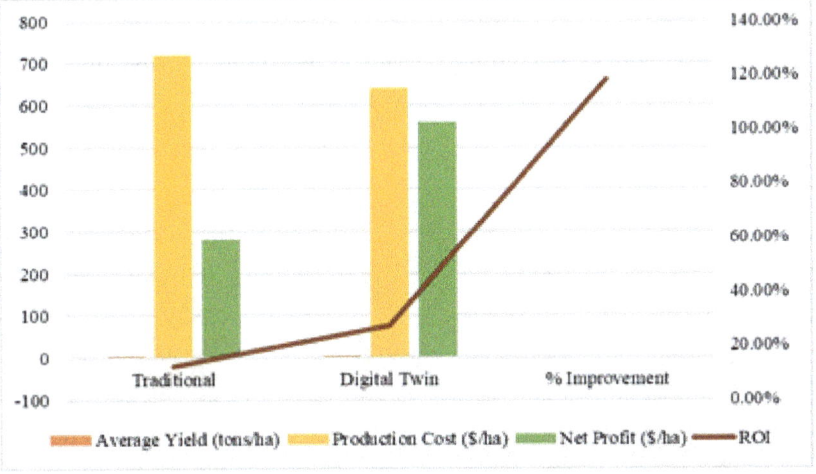

When compared to more traditional approaches, the return on investment (ROI) more than doubles, and the net profit increases by a substantial amount. Even though it is more costly, the benefits far outweigh the cons by a significant margin. According to the ROI curve, which grows to more than 120%, every dollar that is invested in digital twin systems creates considerably larger returns. This is accomplished via the accurate exploitation of resources, the elimination of losses, and the enhancement of input deployment. The information shown in Figure 4 demonstrates without a shadow of a doubt that digital twins are an investment that does not pose any significant financial risks for modern agriculture. Not only do they enhance the outcomes in terms of biology and the environment, but they also have a big favorable influence on the economy. Figure 5 depicts the dramatic impact that digital twin technology has had on agricultural production by using wheat, rice, maize, and soybeans as the categories used in the illustration. The findings unequivocally demonstrate that digital twin systems perform better than traditional approaches across the board for all crop categories. In response to these facts, the output of maize, which is the crop that reacts the most positively, increases by about forty percent, going from 5.2 tons per hectare to over 6.7 tons under the same conditions. The yield of wheat, on the other hand, experiences a rise from around 3.4 tons per hectare to approximately 5 tons per hectare. In a manner comparable to this, the productivity improvements for soybeans and rice range from twenty to thirty-five percent, demonstrating that digital twin technologies are applicable to a wide variety of agricultural contexts rather than being crop-specific. The line that is drawn across the bar graph to represent a percentage increase demonstrates that the yield performance has been steadily improving over the course of time.

Figure 5. Crop Yield Comparison Across Crops – Traditional vs. Digital Twin Methods

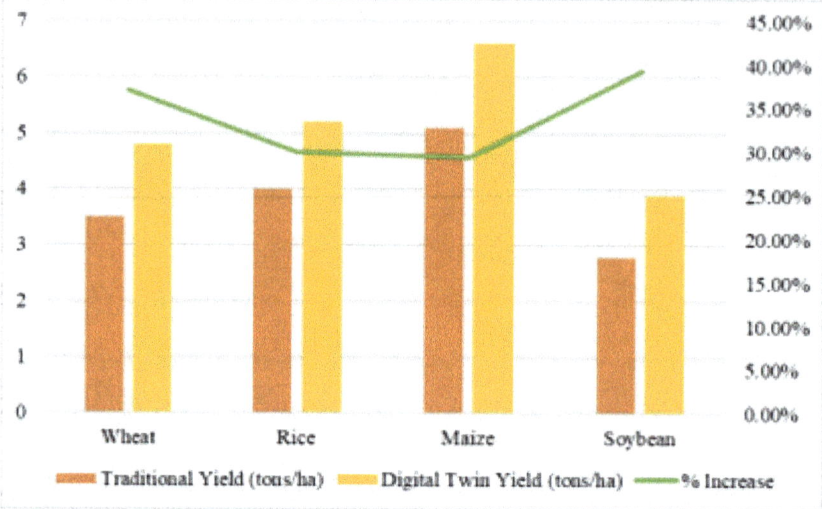

This capability of the digital twin to mimic plant growth, dynamically adjust management strategies, and continually monitor environmental variables has made it possible for these breakthroughs to be achieved. Through the use of these technological advancements, farmers may be able to make more precise assessments, which may result in increased output consistency and decreased variability. It is because of this that farmers have the opportunity to raise their production in a way that is both sustainable and predictable. Figure 6 illustrates a number of significant operational features that were present both before to and after to the implementation of the digital twin technology. Some of benefits include crop output, accuracy of forecasts, decreased prices of inputs, water use, and the amount of time needed to make decisions. The performance has seen a substantial and diversified transformation in recent times. It is possible to increase the accuracy of predictions from 68% to 91%, which would make it possible to make decisions that are substantially timelier and more informed. The crop output also improves, going from 3.6 tons per hectare to 4.8 tons per hectare, which is a 33% improvement between the previous levels. The drop in input costs, which was previously nonexistent, has increased by 22% as a consequence of improvements in the allocation of money for labor, fertilizer, and pesticides.

Figure 6. Key Operational Metrics Before and After Digital Twin Integration

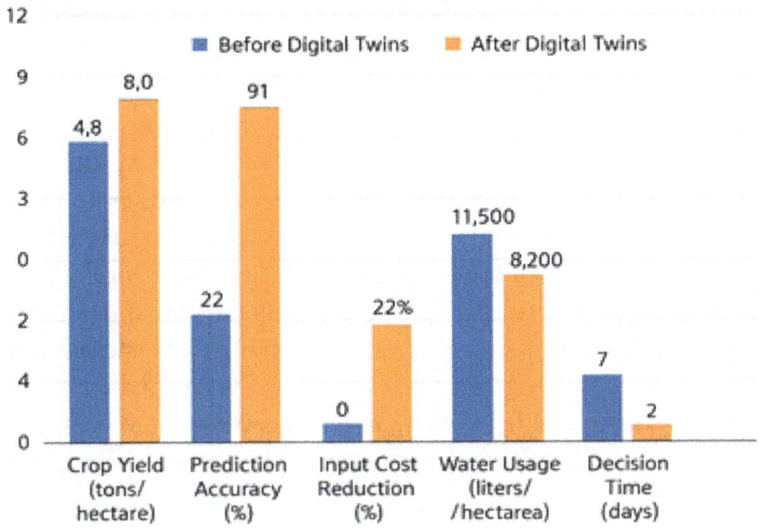

The huge reduction in water use, which went from 11,500 to 8,200 liters per acre, which is equivalent to about 29%, is an example of how digital twins contribute to the sustainability of resources. The decision-making process was reduced from seven days to only two days, which is the single change that stands out as particularly significant. This drop of 71% exemplifies the potential uses of simulation models and real-time data integration by demonstrating how these technologies allow speedier responses to crop stress, pests, weather fluctuations, and changeable field conditions. In conjunction with one another, these enhancements demonstrate how digital twins have the potential to enhance strategic planning and operational execution, so rendering farms more responsive, sustainable, and cost-effective, while simultaneously improving overall productivity.

7. CONCLUSIONS

Research findings show that precision agriculture has developed significantly with the introduction of digital twin technology. By means of real-time data integration, simulation, and predictive modeling, digital twins significantly increase significant agricultural characteristics including crop yield, prediction accuracy, and decision-making efficiency. One may achieve this by means of digital transformation. Digital infrastructure makes this possible. Over a range of agricultural activities including operational responsiveness, water conservation, and input optimization, digital twin systems outperform traditional and sensor-based replacements. Even further the significance of digital twins in offering quantitative and financial advantages is supported by statistical studies including analysis of variance (ANOVA) and correlation testing. Since they enable to boost yield of more than 33 percent, cut water usage of more than 29 percent, and raise return on investment of more than 120 percent, digital twins are not only a technological advancement but also a lucrative answer for modern agriculture. Its scalability and long-term worth are even more shown by its adaptation to a wide range of agricultural settings and crops. Other problems can arise include the initial financial expenditure, the difficulty of system integration, and the demand of higher quality data. Standardizing data formats, developing reasonably cost sensor networks, and future artificial intelligence model improvement should be the main priorities of research to raise adoption particularly in low-resource agricultural regions. Once everything is said and done, digital twins lead the way in Agriculture 4.0, allowing farms to become more intelligent, strong, and sustainable in front of challenges connected to the global security of food supply.

8. REFERENCES

Angin, P., Anisi, M. H., Göksel, F., Gürsoy, C., & Büyükgülcü, A. (2020). AgriLoRa: A digital twin framework for smart agriculture. *Journal of Wireless Mobile Networks, Ubiquitous Computing and Dependable Applications, 11*(4), 77–96.

Howard, D. A., Ma, Z., Aaslyng, J. M., & Jørgensen, B. N. Data Architecture for Digital Twin of Commercial Greenhouse Production. *In Proceedings of the 2020 RIVF International Conference on Computing and Communication Technologies (RIVF),* Ho Chi Minh City, Vietnam, 14–15 October 2020; pp. 1–7.

Hunt, E. R., & Daughtry, C. S. T. (2018). What good are unmanned aircraft systems for agricultural remote sensing and precision agriculture? *International Journal of Remote Sensing, 39*(21), 5345–5376.

Jain, A., Roy, D. K., Ayasrah, F. T., William, P., & Kulkarni, S. "Enhancing Metaverse Cybersecurity With Federated Learning," *Global Congress on Emerging Technologies (GCET-2024)*, Gran Canaria, Spain, 2024, pp. 254-265, doi: DOI: 10.1109/GCET64327.2024.10934363

Javaid, M., Haleem, A., Singh, R. P., & Suman, R. (2022, January). Enhancing smart farming through the applications of Agriculture 4.0 technologies. *International Journal of Intelligent Networks, 3*, 150–164.

Jones, D., Snider, C., Nassehi, A., Yon, J., & Hicks, B. (2020). *Characterising the digital twin: A systematic literature review* (Vol. 29). CIRP Journal of Manufacturing Science and Technology. [Online], Available https://www.sciencedirect.com/science/article/pii/S1755581720300110

Kalyani, Y., Bermeo, N. V., & Collier, R. (2023). Digital twin deployment for smart agriculture in Cloud-Fog-Edge infrastructure. *. *International Journal of Parallel, Emergent and Distributed Systems : IJPEDS, 38*(6), 1–16. DOI: 10.1080/17445760.2023.2235653

Nie, J., Wang, Y., Li, Y., & Chao, X. N. I. E. (2022). Artificial intelligence and digital twins in sustainable agriculture and forestry: A survey. *Turkish Journal of Agriculture and Forestry, 46*(5), 5. DOI: 10.55730/1300-011X.3033

. J. Nyakuri, I. Pierre, J. Viviane, I. Lambert, B. Shadrack, N. Erneste, N. Schadrack, K. Alexis, H. Francois, and Theogene, "Ai based realtime weather condition prediction with optimized agricultural resources," *European Journal of Technology, vol.* 7, 06 2023.

Rojko, A. (2017). Industry 4.0 Concept. *Background and Overview International Journal of Interactive Mobile Technologies, 11*, 77–90.

Sharma, K., ... (2024). Framework for Image Processing-Based Robotic Car for Agricultural Ploughing Using Ensembling Machine Learning Approach. In: Tomar, A., Mishra, S., Sood, Y.R., Kumar, P. (eds) Proceedings of 4th International Conference on Machine Learning, Advances in Computing, Renewable Energy and Communication. MARC 2023. Lecture Notes in Electrical Engineering, vol 1231. Springer, Singapore. https://doi.org/DOI: 10.1007/978-981-97-5227-0_3

William, P., Darwante, N. K., Pawar, A. B., Jawale, M. A., & Verma, A. (2023). Framework for Implementation of Smart Driver Assistance System Using Augmented Reality. In Marriwala, N., Tripathi, C., Jain, S., & Kumar, D. (Eds.), *Mobile Radio Communications and 5G Networks. Lecture Notes in Networks and Systems* (Vol. 588). Springer., DOI: 10.1007/978-981-19-7982-8_30

William, P., Jaiswal, V. K., Shrivastava, A., Bansal, S., Hussein, L., & Singla, A. (2025). "Digital Identity Protection: Safeguarding Personal Data in the Metaverse Learning," *2025 International Conference on Engineering, Technology & Management (ICETM)*, Oakdale, NY, USA, pp. 1-6, doi: DOI: 10.1109/ICETM63734.2025.11051435

William, P., Shrivastava, A., Aswal, U. S., Kumar, I., Gupta, M., & Rao, A. K. (2023). "Framework for Implementation of Android Automation Tool in Agro Business Sector," 2023 4th International Conference on Intelligent Engineering and Management (ICIEM), London, United Kingdom, pp. 1-6, doi: DOI: 10.1109/ICIEM59379.2023.10167328

Yogeesh, N., Girija, D. K., Rashmi, M., & William, P. (2024). Intelligent Irrigation Systems in Agriculture Using Fuzzy Logic Techniques. In Shukla, B., Murthy, B. K., Hasteer, N., Kaur, H., & Van Belle, J. P. (Eds.), *Intelligent IT Solutions for Sustainability in Industry 5.0 Paradigm. ICEIL 2023. Lecture Notes in Electrical Engineering* (Vol. 1185). Springer., DOI: 10.1007/978-981-97-1682-1_25

Chapter 10
Next-Generation Farming:
Digital Twins and Virtual Environments for Sustainable Agriculture

Sandeep Bhatia
https://orcid.org/0009-0005-7854-8265
Jamia Millia Islamia, New Delhi, India

Zainul Abdin Jaffery
https://orcid.org/0000-0002-3502-6599
Jamia Millia Islamia, New Delhi, India

Shabana Mehfuz
Jamia Millia Islamia, New Delhi, India

Neha Goel
https://orcid.org/0000-0002-7189-6305
Raj Kumar Goel Institute of Technology, India

ABSTRACT

- Digital technologies are transforming agriculture, provide innovative solutions for sustainability, efficiency, and resilience in farming. Among these technologies, integration of Digital Twins (DTs) and Virtual Environments (VEs) is revolutionizing agriculture by optimizing agricultural processes, reducing resource consumption, and enhancing productivity. Digital Twins create real-time, data-driven replicas of crops, soil, livestock, and farming infrastructure, enabling continuous monitoring and predictive analytics. These virtual models help farmers make data-informed

DOI: 10.4018/979-8-3373-7077-4.ch010

ght © 2026, IGI Global Scientific Publishing. Copying or distributing in print or electronic forms without written permission of IGI Global Scientific Publishing is prohibited. Use of this chapter to generative artificial intelligence (AI) technologies is expressly prohibited. The publisher reserves all rights to license its use for generative AI training and machine learning model development.

decisions, improving resource efficiency and minimizing environmental impact. Virtual Environments, including Virtual Reality (VR) and Augmented Reality (AR), complement Digital Twins by providing immersive simulations for precision farming, training, and farm management.

1. INTRODUCTION

Sustainable agriculture has been practiced for ages in every country. Agriculture is a science and an art that involves the cultivation of plants. It has a major impact on everyone's capacity to live and conduct daily life worldwide. Additionally, agriculture contributes to the economic expansion of humankind's complete natural life. In the age of technological advancement, contemporary agriculture, or smart farming, is made possible by new digital technology. In the realm of agriculture, smart technology greatly supports and enables the latest Internet of Things (IoT) technologies. Today's challenges cannot be solved without technology, and many sectors are using the newest innovations to automate their processes. By using efficient techniques and tools to manage crops, cattle, and land, IoT is revolutionizing agriculture. Consequently, there has been a notable reduction in waste production and an exceptional rise in output, Purcell et al. (2023).

Smart agriculture is a relatively new concept. Most farmers and agricultural experts are not familiar with this concept. Smart agriculture is the use of intelligent technologies to control and manage agricultural areas and livestock. Drones, sensors, actuators, surveillance cameras, and automated machinery are a few examples of these technologies. These IoT technologies are very much helpful for the farmers in the agriculture sector: (i) Smart agriculture is a new idea since IOT sensors can collect information about agricultural fields and then operate in response to user input. Insect and crop conditions are also tracked and recorded using sensors. (ii) Drones are employed to observe animals and keep an eye on routine duties. (iii) Automatic water pumping systems that irrigate crops when they need it. (iv) Machines and devices that perform route operations and ensure that systems are functioning correctly. Increasing the quantity and quality of agricultural products while taking cost and energy use into account is the aim of smart agriculture, Nasirahmadi et al. (2022).

In terms of social and economic advantages, the Internet of Things (IoT) has a lot to offer emerging and developing nations. It has been widely used in agricultural management to help farmers monitor and manage their crops more successfully and efficiently.

With millions of people in India depending on agriculture for their livelihoods and the nation's economy, the use of Internet of Things (IoT) technology for crop management may prove advantageous. Encouraging the food supply chain to be optimized can not only greatly boost economic growth but also aid in the global eradication of poverty and hunger, Vitali et al. (2023).

Figure 1. Digital twin in IoT based Agriculture Applications, Verdouw et al. (2021)

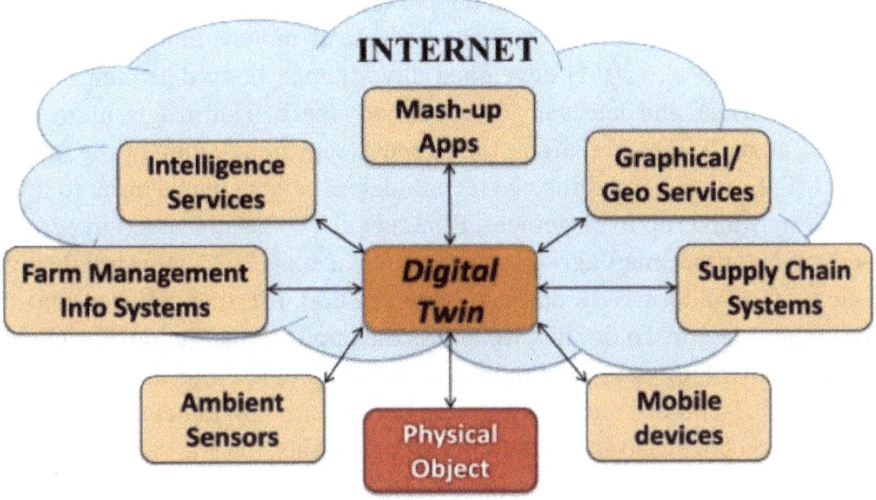

As shown in Fig 1, the network of linked devices with sensors, software, and other technologies integrated in them that allow them to gather and share data is known as the Internet of Things (IoT) and digital tins are the key pillars. Verdouw et al. (2013), Real-time data collection, analysis, and decision-making are made possible by this seamless connectivity between physical devices. Although IoT has applications in a number of industries, including smart cities, transportation, and healthcare, its influence on agriculture has been especially revolutionary.

The incorporation of cutting-edge digital technologies is causing a revolutionary change in the agriculture industry. Sustainable, effective, and resilient farming methods are more important than ever as concerns about resource scarcity, climate change, and food security increase. Digital Twins (DTs) and Virtual Environments (VEs) are two of the major breakthroughs driving this shift, changing the way that farming is perceived and carried out. This chapter explores how DTs and VEs are transforming agriculture by helping farmers increase productivity, cut waste, and improve operations.

2. LITERATURE REVIEW

Hammad Shahab, et.al. (2024) highlights the technique which can reduce the waste of expensive fertilizer, water, and electricity, among other things. The device will automatically maintain the soil's moisture content and fertility at ideal levels. in order to achieve the most accurate and ideal crop growth results. This approach will reduce labor costs, save a significant amount of money, and make farming prosperous. Bhatia et al. (2011) developed alive human body detection that can roam around farm land and detect any object or body that can harm agriculture land.

Vitali, G., et.al (2021) summarizes the elements and architecture of the Internet of Things (IoT) from an agronomic and technological perspective in order to investigate how it may aid crop management. Bhatia et al. (2023) discussed integration of IoT and WSN for the smart agriculture. Quy, V.K., et.al (2022), this article gives an overview of IoT alternatives and demonstrates how IoT may be used into the smart agriculture sector. To do this, we assess the applications, research schedule, and architecture (IoT devices, communication technologies, big data processing and storage) of IoT-enabled smart agriculture ecosystems. Bhatia et al. (2023) discussed WSN protocols used in precision farming for the smart agriculture. Cesco et al. (2023) discussed digital twins with respect to smart agriculture. Rohit Vaish, et.al. (2024) highlighted to optimize plant health and yield, the emphasized research aims to create a remote system that can continuously monitor soil moisture. Bhatia et al. (2023) discussed 5G to 7G revolution and their role in precision farming. for the smart agriculture.

Prem Rajak, et.al (2023) suggested using Internet of Things (IoT) smart sensors to track soil composition, temperature, moisture, and humidity all of which are critical for crop growth. Bhatia et al. (2023) discussed IoT role and its reliability for smart agriculture applications.

Sandya De Alwis, et.al (2022) discussed IoT hardware for the smart agriculture. Bhatia et al. (2024) discussed about machine learning (ML)-based smart weather prediction that can be applied to smart agricultural precision farming. Ranjitha et.al. (2022) stated that managing solid waste is a crucial problem in cities. An unsterilized atmosphere caused by an overflowing garbage poses a number of health risks, including sore throats, lungs, and noses. Effective waste management is necessary to get past these problems. The Internet of Things, or IoT, has had a big impact. The smart bin, which consists of an Arduino microcontroller, an ultra-sensor, and a rain sensor, is represented by the proposed work. Once the dustbin is full, Aurdino is configured to provide a notification via the Bylank app. Dhillon et al. (2024), Goel et al. (2023) & Jeet et al. (2015) discussed role of IoT and its energy consumption of WSN nodes. Kumar et al. (2024) discussed NLP based sentiment analysis and it prediction techniques. Ganai et al. (2022) discussed IoT and its cloud security required in smart agriculture.

Kalogiannidis, S., et.al. (2024), according to research, there is either a positive correlation or a substantial effect between sustainable agricultural productivity and the functions of mechanical, chemical, and biological crop protection measures. This work also explains regression and descriptive analysis of the present environmental, political, and economic issues. The agriculture industry should enhance and further implement current technology, expand the use of integrated pest-management techniques, and completely incorporate innovations into agricultural management and production in order to handle these problems. Bhatia et al. (2023) discussed ML used in precision farming for the smart agriculture.

Balasubramaniyan et.al. (2021) Using Internet of Things (IoT) technologies, a smart farming system was proposed. The MAC, SMAC, and TMAC protocols used in precision farming for smart agriculture were covered by Bhatia et al. (2023).

Li, C.-H. G., et.al. (2021), discussed deep learning methods for disease detection of crop leaves. IoT smart apps for precision farming were covered by Bhatia et al. (2024) as part of smart agriculture.

Kulkarni, S. A., et.al (2020) The primary problem being addressed in this work is the development of an effective water monitoring system based on level sensors that keeps track of the water level within homes and other domestic places. Using depth sensors, the suggested system will measure the water level and confirm the predetermined threshold.

Gogul Dev, N. S., et.al (2019) This system will provide farmers with a complete technical solution to safeguard their crops from wild animals and provide them with information to assist them boost productivity, all thanks to the Internet of Things (IOT). Animals are identified using TensorFlow image processing techniques after being detected by PIR sensors and cameras. The system's computing unit is the Raspberry Pi, and buzzers are used to emit ultrasonic frequencies. Bhatia et al. (2023) discussed WSN heterogeneous nodes used in precision farming for the smart agriculture.

Wang, S.-W., et.al. (2018), suggested a continuous water level sensor that uses floating pipes and load cells. Many different fields have made extensive use of the liquid-level sensing techniques currently employed in monitoring systems. The float gauge and pressure gauge used in earlier equipment have been replaced by the ultrasonic gauge and radar gauge. Bhatia et al. (2023) discussed 5G to 7G revolution and its role in precision farming. for the smart agriculture.

Sushanth et.al., (2018) suggested developing a smart agriculture system that utilizes state-of-the-art technologies such as the Internet of Things, Arduino, and wireless sensor networks. The recommended automatic intelligent farming system, tracks temperature, moisture, humidity, and even the movement of animals that could damage crops in agricultural fields using sensors built on an Arduino board. The system sends an SMS to farmers' smartphones in the event of a discrepancy. Bhatia et al. (2023) shows the ML use in all applications include precision farming for the smart agriculture. R. Balathandapani, et.al. (2015) Using an auto roof that covers the entire field to protect the crops, this technique is suggested to conserve rainwater and protect the crops from heavy rain. It is operated by a rain sensor and a soil moisture sensor.

Bhatia et al. (2023) discussed ML use in driver drowsy detection used in precision farming for the smart agriculture. Oladunmoye M, et.al. (2014) proposed a sliding door that operates automatically. An infrared sensor-based system was created. It opens and closes doors at a public building's entrance using a sensor, a control unit, and a drive unit. With the creation of a basic circuit model and their fundamental operations, this research project offers information on how an automatic door system operates. Elahi, 2014 suggested a vehicle One of the most crucial driving abilities is controlling the windshield. Sensors and microcontrollers can be used to automate the windshield for smooth driving. A water level sensor regulates the wiper motor by measuring the amount of water or precipitation. In order to allow drivers to focus entirely on the road, even in the summer when it may be raining or dirty, the suggested method aims to make the current system more compassionate. The purpose of this technology is to increase human comfort and flexibility by providing a comprehensive windshield regulating mechanism.

Zhaolong Xuan, et.al. (2012) offered a monitoring microsystem that is intended to be installed in a small shelter system used for munitions transportation. It can give the customer the information they need to analyze the mishap or assess the complex ammunition's capabilities. A few major issues with the design of the microsystem are explored, such as power consumption and multi-dimension coupling. O'Flynn, B., et.al., (2007) suggested a multisensory water quality monitoring system called "Smart Coast." The goal of this system is to offer a platform that can satisfy the Water Framework Directive Smart Coast monitoring requirements, Navarro et al. (2020)

Bhatia et al. (2024) discussed IoT integration with other applications in cloud computing.

Bhatia et al. (2023) discussed WSN heterogeneous nodes used in precision farming for the smart agriculture. Bhatia et al. (2023) discussed 5G to 7G revolution and its role in precision farming. for the smart agriculture. Bhatia et al. (2023) shows the ML use in all applications include precision farming for the smart agriculture. Bhatia et al. (2023) discussed ML use in driver drowsy detection used in precision farming for the smart agriculture. Jarrett et al. (2021) discussed role of AI in digital forensic. Yew et.al., (2022) suggested a Illinois BIPA in digital forensic. Mallikarjuna et al. (2023) et al. discussed 5G to 7G applications in cloud computing. Jung et al. (2024) discussed framework for biometric data. Paolucci et al. (2023) discussed safeguards and standard in digital forensic. Drowsy detection and finger print recognition for driver discussed by Bhatia et a. (2023). Gautam et al. (2023) discussed smart traffic light system that used cloud computing. Bhatia et al. (2023) shows the ML based prediction system in healthcare and Goel et al. (2023) used ML in solar energy. Madan et al. (2023) used latest technologies in fire safety in remote areas. Srinivasan et al (2023) discussed python use in ML and Singh et al. (2024) shows the recent advancement in healthcare. Table 1 indicates literature review implications.

Inferences from the Literature Review

Table 1. shows the literature review implications

Concept Aspect	Reference	Discussion
IoT and Smart Farming	Anamul et al. (2024) Prem et al. (2023) Sandya et al. (2022) Sushanth et al. (2018)	• Discussed IoT's role in reducing resource wastage and improving productivity by 20% through real-time analytics. • Highlighted the use of smart sensors for monitoring soil conditions, which significantly enhanced yield quality.
Automated Systems	Gogul dev et al. (2019) Wang et al. (2018) Balathandapani et al. (2015) Oladunmoye et al. (2014) Ariffin et al. (2020)	• Explored the application of IoT for predictive analytics in farming, showing improved crop health monitoring and decision-making. • Proposed an embedded system for rainwater management and crop shelter automation.
Sustainability in IoT Farming	Hammad et al. (2024) Kalogiannidis et al. (2024) Sandya et l (2022) Ahmed et al. (2023)	• Studied the integration of IoT with biological and mechanical crop protection methods to enhance sustainability. • Examined IoT-enabled smart agriculture ecosystems, focusing on architecture and data-driven farming.
Emerging Trends and Applications	Kalaiselvi et al. (2024) Valarmathi et al. (2021) Govindaraj et al. (2023)	• Proposed IoT systems for protecting crops from wild animals using PIR sensors and image recognition • Discussed IoT-based systems for automated irrigation and pest detection, highlighting cost and labor savings.

3. BACKGROUND

The following items can be added to address the lack of background data and a thorough literature review:

- Insufficient or lack of funds
- Agricultural promoting
- Soil erosion
- Issues with irrigation
- Infrastructure deficiencies in the agricultural industry

It is difficult for farmers to get affordable, high-quality seeds and insecticides. In contrast to other nations like China or the USA, where all the amenities are at their fingertips, they must also spend a significant amount of money on fertilizers and other inputs like irrigation systems, etc.

4. ROLE OF TECHNOLOGIES IN DIGITAL TWINS AND VIRTUAL ENVIRONMENTS FOR SUSTAINABLE AGRICULTURE

As shown in fig 2., technologies such as IoT, AI, machine learning, remote sensing, cloud computing, and AR/VR play a crucial role in building effective Digital Twins (DTs) and Virtual Environments (VEs) to enable sustainable agriculture. These technologies help simulate, monitor, and optimize farming activities in real-time, reducing resource usage and environmental impact. Some main points discussed below:

- IoT Sensors: Collect real-time data from fields (soil moisture, temperature, pH, etc.) and livestock (health, movement), which feeds into DTs for accurate modeling.
- Artificial Intelligence & Machine Learning: Analyze complex data sets to predict crop yields, detect disease early, and recommend precision interventions.
- Remote Sensing & Drones: Provide aerial data (NDVI, thermal imaging) for virtual environment mapping and decision support.
- Cloud & Edge Computing: Store, process, and analyze vast agricultural data in real-time to support scalable DT applications.
- AR/VR Technologies: Create immersive virtual environments for training, simulation, and planning agricultural activities.
- GIS & Spatial Data Analysis: Help visualize and model land use, weather patterns, and irrigation requirements within VEs.

Together, these technologies enhance decision-making, reduce resource waste (like water and fertilizer), improve crop and livestock health, and support climate-resilient farming—contributing to a smarter, more sustainable agricultural ecosystem.

AI and Machine Learning in Predictive Analytics

The use of machine learning (ML) and artificial intelligence (AI) to IoT device data is changing the agricultural industry. These technologies use massive volumes of data produced by sensors, drones, and other devices to analyze it in order to find trends and forecast the development of diseases, insect outbreaks, and agricultural performance.

Figure 2. Role of technologies in digital twins for sustainable agriculture

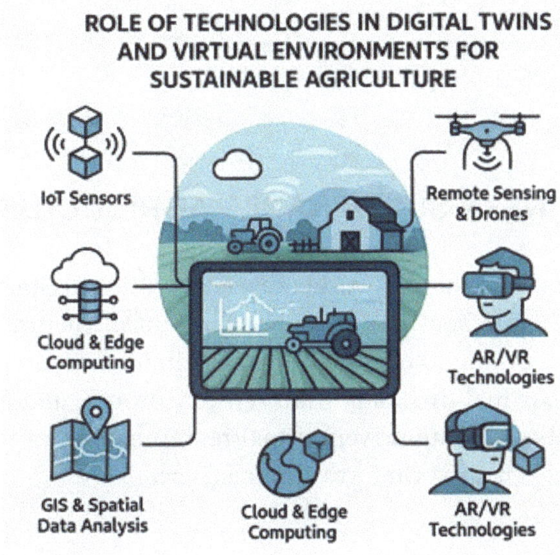

Integration with Other Farm Management Systems: Figure 3 illustrates the integration model.

Figure 3. Sustainable and Smart IoT-Based Irrigation

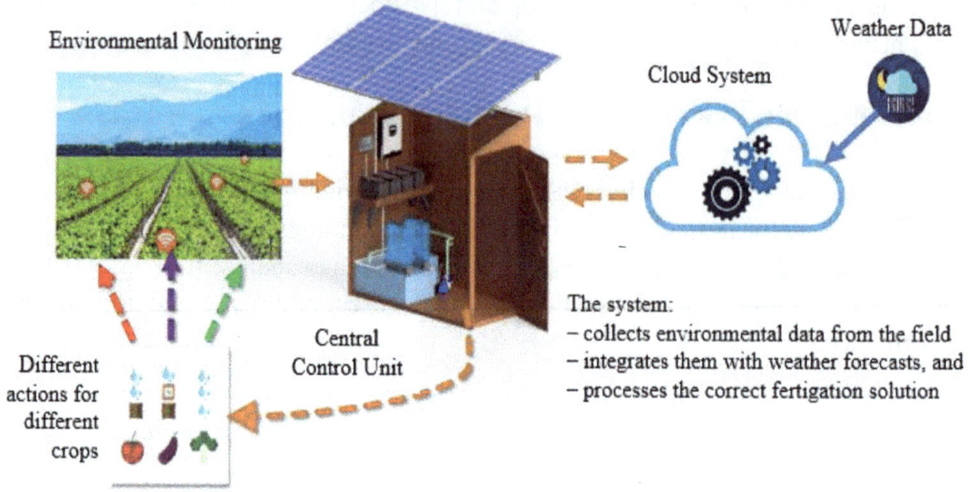

5. DIGITAL TWINS IN SUSTAINABLE AGRICULTURE

Virtual representations of real-world objects are called digital twins. Crops, soil systems, livestock, equipment, and complete farm infrastructures are examples of these in agriculture. Continuous monitoring and simulations are made possible by feeding these replicas real-time data from sensors, drones, and Internet of Things devices. It is a real-time, data-driven virtual representation of the actual farm. It mimics and simulates real-world farm circumstances.

Figure 4. Technology role in smart farming

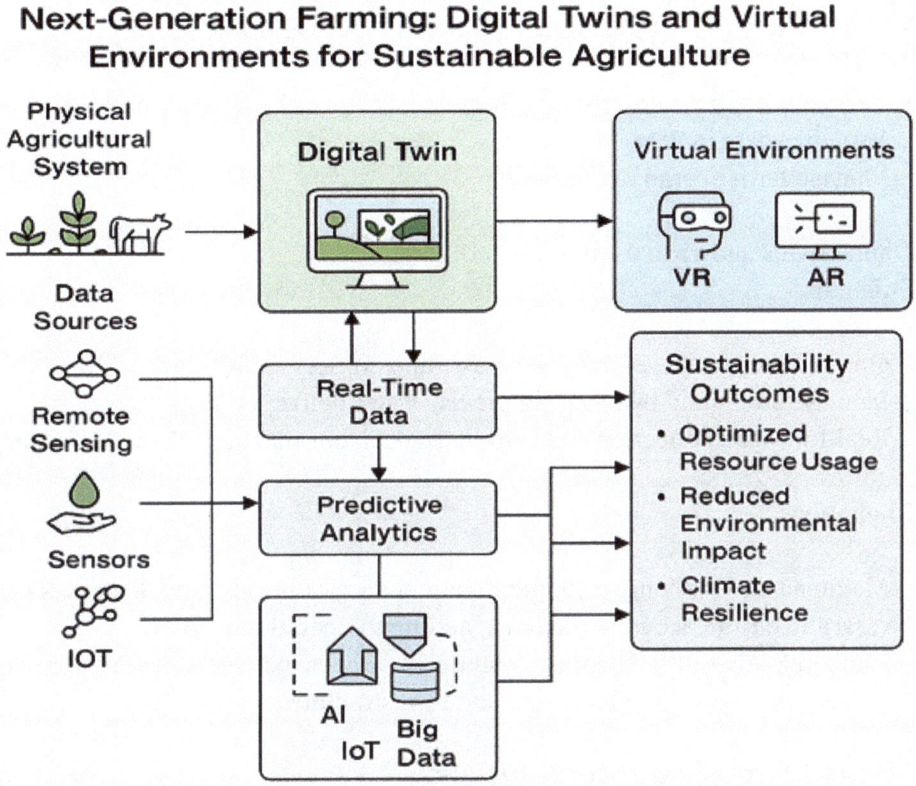

Figure 4 shows how cutting-edge technologies work together in contemporary precision farming.

Applications

- **Crop Management:** DTs simulate plant growth, forecast yields, and assess disease risk.
- **Soil Monitoring:** Real-time moisture, pH, and nutrient levels enable optimized fertilization.
- **Livestock Management:** Monitoring animal health, movement, and productivity helps improve welfare and output.
- **Infrastructure Management:** Predictive maintenance for machinery and irrigation systems.

6. VIRTUAL ENVIRONMENTS

Virtual Reality (VR) and Augmented Reality (AR) allow users to:

- Interact with digital twins
- Visualize data in 3D
- Manage farm operations virtually

Digital twins and virtual environment:
Strengths:

- Optimize water usage, reducing waste and conserving resources.
- Increase crop yield by ensuring precise water delivery.
- Real-time data collection and automated decision-making enhance efficiency.

Limitations:

- Maintenance Challenges: Regular maintenance of sensors and devices is necessary to ensure accurate readings, adding to operational costs.
- Limited Scalability: Adapting systems to large-scale farms may require significant investments in hardware and infrastructure.

IoT-Based Crop Security Solutions

Strengths:

- Protect crops from threats such as pests, diseases, and unauthorized intrusions.
- Provide real-time alerts, enabling timely action and minimizing losses.
- Enhance food security by reducing the risk of crop failure.

Limitations:

- Energy Consumption: Continuous monitoring systems may have high energy demands, requiring sustainable power sources, particularly in off-grid areas.
- Data Privacy Concerns: The use of cameras and sensors in fields raises concerns about data security and unauthorized data access.

7. ARCHITECTURE OF DIGITAL TWINS AND VIRTUAL ENVIRONMENTS FOR SUSTAINABLE AGRICULTURE

The proposed architecture integrates IoT, AI, and ML into a single, integrated system for efficient carbon monitoring, analysis, and optimization. It is made up of interconnected parts that cooperate to gather data, process it, and produce useful insights for cutting carbon emissions as discussed in fig 5.

Figure 5. Architecture of Digital Twins and Virtual Environments for Sustainable Agriculture

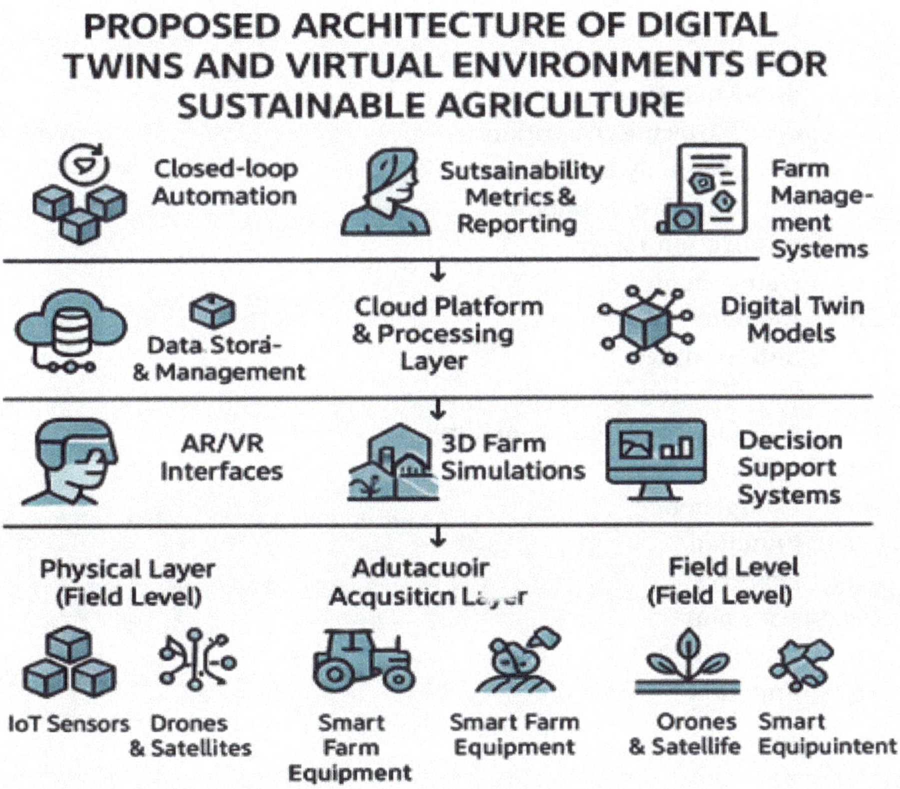

Architecture Components

1. **Physical Layer (IoT Devices)**
   - **Sensors**: Overview of these sensors shows in Table 2.
   - **Devices**

- **Function**
- **Connectivity**

2. **Edge Layer (Pre-Processing)**
   - **Edge Devices**
   - **Function**
   - **Output**
3. **Data Aggregation Layer**
   - **Component**
   - **Function**
   - **Features**
4. **AI Processing Layer**
   - **Component**
   - **Function**
   - **Sub-Modules**:
     - **Pattern Recognition**
     - **Anomaly Detection**
     - **Simulation Engine**
5. **ML Optimization Layer**
   - **Component**
   - **Function**.
   - **Sub-Modules**
     - **Predictive Models**
     - **Optimization Algorithms**
6. **Application Layer**
   - **Component**
   - **Function**
   - **Features**
7. **Feedback Loop**
   - **Component**
   - **Function**

8. SENSORS AND ITS CLASSIFICATION USED IN SMART AGRICULTURE

Table 2. Various sensor types are employed in different agricultural sectors

Sensor Type	Working Principle	Usage	Accuracy	Precision	Sensitivity	Specifications
Soil Moisture Sensors	Capacitance methods (detecting changes in dielectric constant of soil) or resistive methods.	Real-time soil moisture data for optimizing irrigation and preventing over- or under-watering.	High	High	High	Range: 0-100% VWC, Accuracy: ±3%, Response time: <1 second
Temperature and Humidity Sensors	Thermistors or RTDs for temperature; capacitive or resistive hygrometers for humidity.	Monitoring greenhouse environments to maintain optimal conditions for plant growth.	High	High	Moderate	Accuracy: ±0.5°C for temperature range -40°C to 85°C and ±2% for humidity range 0-100% RH
Nutrient Sensors	Ion-selective electrodes (ISE) or optical methods (light absorption or fluorescence).	Monitoring nutrient levels for precise fertilization and minimizing waste.	Very high	High	High	Range: 0-10 mM for ions, Accuracy: ±5%, Response time: 2-10 seconds
Crop Health Sensors	Optical methods (multispectral/hyperspectral imaging) to analyze light reflected from plants.	Early identification of diseases, insects and nutrient deficits allows for targeted response.	High	Very high	Very high	Wavelength: 400-1000 nm (multispectral), Resolution: 1-2 nm, Response time: <1 second
Weather Sensors	Anemometers (wind speed), barometers (pressure), rain gauges (precipitation measurement).	Provides critical weather data such as wind, precipitation, and pressure for farming decisions.	Moderate	Moderate	High	0–40 m/s wind speed, accuracy ±0.1 m/s; 300–1100 hPa pressure range, accuracy ±1 hPa

9. FUTURE SCOPE

Future advancements in IoT-based smart agriculture monitoring systems could include the following:

Combining AI and ML Techniques: AI and ML can be linked with IoT-based smart agriculture monitoring systems to facilitate decision-making and predictive analytics. AI systems, for instance, are able to evaluate data from soil sensors and forecast, using past data, the best time to plant various crops.
Advancement with Edge Computing:Edge computing handles data at the network's edge, nearer the point of generation, as opposed to transferring it to a central cloud. Consequently, decision-making is sped up and latency is reduced. In the context of smart crop monitoring, edge computing can be used to process data from farm sensors and cameras, allowing farmers to make decisions in real time without relying on the cloud.
Integration with Blockchain: By integrating blockchain technology with IoT-based smart agricultural monitoring systems, safe and transparent data sharing between various agriculture value chain participants is made possible. In order to guarantee that they receive a fair price for their produce, producers might provide buyers with information on crop yields and quality.
Rise the Drone Usage: The Drone usage is on the rise because farmers can quickly decide on fertilization and irrigation by using real-time data that drones can gather on crop conditions and soil moisture levels. Autonomous drones may now be programmed by farmers to fly over their crops and collect data without requiring human intervention.

These developments will eventually enable farmers to increase crop yields, make data-driven choices, and optimize resource use, resulting in a more sustainable and effective agriculture industry.

CONCLUSION

Proposed system offer numerous advantages, including improved resource efficiency, optimized irrigation and fertilization, enhanced crop health monitoring, and increased yield predictability. By providing real-time insights into soil conditions and predicting future trends, farmers can make more informed decisions that not only boost productivity but also reduce environmental impact. However, challenges such as high initial costs, technical expertise requirements, and reliance on stable internet connectivity pose barriers to widespread adoption, particularly for smaller or rural farming operations. Despite these limitations, many of the current IoT and machine learning methods—such as soil moisture sensing, precision irrigation, and crop health prediction—are already proving to be highly effective in modern farming practices. As technology continues to advance, these tools are likely to become more accessible and cost-effective, paving the way for broader implementation and transforming the agricultural industry into a more sustainable and efficient sector.

A new horizon in sustainable agriculture is represented by virtual environments and digital twins. These technologies facilitate more intelligent, effective, and ecologically responsible farming by enabling real-time monitoring, predictive modeling, and immersive visualization. Despite certain obstacles, there is no denying their capacity to make agriculture a more resilient and sustainable industry. This chapter emphasizes how important it is to integrate DTs and VEs in order to shape food systems going forward.

CONSENT FOR PUBLICATON

None Declare

CONFLICT OF INTEREST

None Declare

ACKNOWLEDGEMENT

None Declare

REFERENCES

Ahmed, L. J., Dhanasekar, S., Govindaraj, V., & Ezhilazhagan, C. (2023). Crop Management System. *Advanced Technologies for Smart Agriculture*, (Dec), 55–79. DOI: 10.1201/9781032628745-3

Ariffin, M. A. M., Ramli, M. I., Amin, M. N. M., Ismail, M., Zainol, Z., Ahmad, N. D., & Jamil, N. (2020). ''Automatic Climate Control for Mushroom Cultivation Using IoT Approach,'' *In Proceedings of the 2020 IEEE 10th International Conference on System Engineering and Technology (ICSET),* Shah Alam, Malaysia, pp. 123–128.

Balasubramaniyan, M., & Navaneethan, C. (2021), "Applications of Internet of Things for smart farming – A survey", *Materials today proceedings, 12th National Conference on Recent Advancements in Biomedical Engineering*, Vol. 47, Part 1, 2021, Pages 18-24, https://doi.org/DOI: 10.1016/j.matpr.2021.03.480

Balathandapani, R., Boopathi, D., Jotheeshwaran, S., Arundeva, G., & Saranya, C. (2015). Automatic rain water and crop saving system using embedded technology. International Journal of Science, Engineering and Technology Research (IJSETR), 4(3).

Bhatia, S., Arya, C., Verma, S., Gautam, D., Naib, B. B., & Kumar, A. (2023). Predict success of a Zomato restaurant using machine learning. In *2023 World Conference on Communication & Computing (WCONF)* (pp. 1–7). IEEE. https://doi.org/DOI: 10.1109/WCONF58270.2023.10235064

Bhatia, S., Bhadauria, R. V. S., Mehrotra, K., Malik, J., Goel, A. K., & Naib, B. B. (2023). A comparative analysis and design criteria of MAC, SMAC, and TMAC protocols in wireless sensor network. In *2023 14th International Conference on Computing Communication and Networking Technologies (ICCCNT)* (pp. 1–6). IEEE. https://doi.org/DOI: 10.1109/ICCCNT56998.2023.10306758

Bhatia, S., Dhillon, H. S., & Kumar, N. (2011). Alive human body detection system using an autonomous mobile rescue robot. In *2011 Annual IEEE India Conference* (pp. 1–5). IEEE. https://doi.org/DOI: 10.1109/INDCON.2011.6139388

Bhatia, S., Gautam, D., Kumar, S., & Verma, S. (2023). Automatic seat identification system in smart transport using IoT and image processing. In *2023 3rd International Conference on Intelligent Communication and Computational Techniques (ICCT)* (pp. 1–6). IEEE. https://doi.org/DOI: 10.1109/ICCT56969.2023.10075664

Bhatia, S., Goel, A. K., Naib, B. B., Singh, K., Yadav, M., & Saini, A. (2023). Diabetes prediction using machine learning. In *2023 World Conference on Communication & Computing (WCONF)* (pp. 1–6). IEEE. https://doi.org/DOI: 10.1109/WCONF58270.2023.10235187

Bhatia, S., Goel, N., Ahlawat, V., Naib, B. B., & Singh, K. (2023). A comprehensive review of IoT reliability and its measures: Perspective analysis. In *Handbook of research on machine learning-enabled IoT for smart applications across industries* (pp. 365–384). IGI Global.

Bhatia, S., Goel, N., Dandeva, D., Sharma, N., Mallikarjun, B., & Naib, B. B. (2024). Detection of posture using MS-IoT-based solution to provide timely treatment. In *Using machine learning to detect emotions and predict human psychology* (pp. 209–234). IGI Global.

Bhatia, S., Goel, N., Jaffery, Z. A., & Mehfuz, S. (2023). Integration of WSN and IoT: Wireless networks architecture and protocols—A way to smart agriculture. In *Handbook of research on machine learning-enabled IoT for smart applications across industries* (pp. 435–455). IGI Global.

Bhatia, S., Jaffery, Z. A., & Mehfuz, S. (2023). A comparative study of wireless communication protocols for use in smart farming framework development. In *2023 3rd International Conference on Intelligent Communication and Computational Techniques (ICCT)* (pp. 1–7). IEEE. https://doi.org/DOI: 10.1109/ICCT56969.2023.10075696

Bhatia, S., Jaffery, Z. A., & Mehfuz, S. (2023). Development and analysis of IoT based smart agriculture system for heterogenous nodes. In *2023 International Conference on Recent Advances in Electrical, Electronics & Digital Healthcare Technologies (REEDCON)* (pp. 62–67). IEEE. https://doi.org/DOI: 10.1109/REEDCON57544.2023.10150466

Bhatia, S., Mallikarjun, B., Naib, B. B., & Goel, A. K. (2023). Algo-indicator based trading in stock market. In *2023 4th International Conference on Smart Electronics and Communication (ICOSEC)* (pp. 1730–1734). IEEE. https://doi.org/DOI: 10.1109/ICOSEC58147.2023.10275858

Bhatia, S., Mallikarjuna, B., Gautam, D., Gupta, U., Kumar, S., & Verma, S. (2023). The future IoT: The current generation 5G and next generation 6G and 7G technologies. In *2023 International Conference on Device Intelligence, Computing and Communication Technologies (DICCT)* (pp. 212–217). IEEE. https://doi.org/DOI: 10.1109/DICCT56244.2023.10110066

Bhatia, S., Naib, B. B., Goel, A. K., Kumar, L., & Singh, K. (2023). Drowsy driver alert system using OpenCV. In *2023 3rd International Conference on Pervasive Computing and Social Networking (ICPCSN)* (pp. 614–619). IEEE. https://doi.org/ DOI: 10.1109/ICPCSN58827.2023.00108

Bhatia, S., Naib, B. B., Goel, N., Agarwal, B., Mallikarjun, B., & Arvindhan, M. (2024). A new era in weather forecasting: Harnessing the potential of machine learning. In *2024 IEEE International Conference on Computing, Power and Communication Technologies (IC2PCT)* (pp. 1355–1361). IEEE. https://doi.org/DOI: 10.1109/IC2PCT60090.2024.10486245

Cesco, S., Sambo, P., Borin, M., Basso, B., Orzes, G., & Mazzetto, F. (2023). Smart agriculture and digital twins: Applications and challenges in a vision of sustainability. *European Journal of Agronomy*, *146*, 126809.

De Alwis, S., Hou, Z., Zhang, Y., & Na, M. H. (2022, June). "A survey on smart farming data, applications and techniques. *Computers in Industry*, *138*. Advance online publication. DOI: 10.1016/j.compind.2022.103624

Dhillon, H. S., Chawla, P., Kumar, K., & Bhatia, S. (2024). Study the effect of hybrid energy harvesting wireless body sensor nodes using multi-dimensional approach in simulation. In *2024 IEEE 5th India Council International Subsections Conference (INDISCON)* (pp. 1–5). IEEE. https://doi.org/DOI: 10.1109/INDISCON62179.2024.10744340

Elahi, A. F., & Rahman, M. S. (2014, December). Intelligent windshield for automotive vehicles. In 2014 17th international conference on computer and information technology (ICCIT) (pp. 392-396). IEEE.

Ganai, P. T., Bag, A., Sable, A., Abdullah, K. H., Bhatia, S., & Pant, B. (2022). A detailed investigation of implementation of Internet of Things (IoT) in cyber security in healthcare sector. In *2022 2nd International Conference on Advance Computing and Innovative Technologies in Engineering (ICACITE)* (pp. 1571–1575). IEEE. https://doi.org/DOI: 10.1109/ICACITE53722.2022.9823887

Gautam, D., Bhatia, S., Goel, N., & Mallikaijuna, B., H S, G., & Naib, B. B. (2023). Development of IoT enabled framework for LPG gas leakage detection and weight monitoring system. In *2023 International Conference on Device Intelligence, Computing and Communication Technologies (DICCT)* (pp. 182–187). IEEE. https://doi.org/DOI: 10.1109/DICCT56244.2023.10110294

Gautam, D., Bhatia, S., Neha, P., & Naib, B. B. (2023). IoT based automatic traffic light and speed breaker system. In *2023 3rd International Conference on Innovative Sustainable Computational Technologies (CISCT)* (pp. 1–6). IEEE. https://doi.org/DOI: 10.1109/CISCT57197.2023.10351445

Goel, N., Bhatia, S., & Rana, P. (2023). Secure smart devices control with low power consumption. In *2023 3rd International Conference on Advancement in Electronics & Communication Engineering (AECE)* (pp. 83–87). IEEE.

Gogul Dev, N. S., Sreenesh, K. S., & Binu, P. K. " IoT Based Automated Crop Protection System," *2019 2nd International Conference on Intelligent Computing, Instrumentation and Control Technologies (ICICICT)*. doi:, 2019.DOI: 10.1109/icicict46008.2019.899

Govindaraj, V., Dhanasekar, S., Martinsagayam, K., Pandey, D., Pandey, B. K., & Nassa, V. K. (2023, January). Low-power test pattern generator using modified LFSR. *Aerospace Systems*, *7*(1), 67–74. DOI: 10.1007/s42401-022-00191-5

Jeet, V., Dhillon, H. S., & Bhatia, S. (2015). Radio frequency home appliance control based on head tracking and voice control for disabled person. In *2015 Fifth International Conference on Communication Systems and Network Technologies* (pp. 559–563). IEEE. https://doi.org/DOI: 10.1109/CSNT.2015.189

Kalaiselvi, K., Anand, A. J., Tanwar, P., & Raza, H. (Eds.). (2024). *Advanced technologies for smart agriculture*. CRC Press.

Kalogiannidis, S., Kalfas, D., Chatzitheodoridis, F., & Papaevangelou, O. (2024). Role of Crop-Protection Technologies in Sustainable Agricultural Productivity and Management. *Land (Basel)*, *11*. Advance online publication. DOI: 10.3390/land11101680

Kulkarni, S. A., Raikar, V. D., Rahul, B. K., Rakshitha, L. V., Sharanya, K., & Jha, V. (2020). " Intelligent Water Level Monitoring System Using IoT," *IEEE International Symposium on Sustainable Energy, Signal Processing and Cyber Security (iSSSC)*. doi:DOI: 10.1109/isssc50941.2020.93588,2020

Kumar, D., Bhatia, S., Dhillon, H. S., & Goel, A. K. (2024). NLP-based sentiment analysis using deep learning methods. In *2024 1st International Conference on Advances in Computing, Communication and Networking (ICAC2N)* (pp. 1873–1878). IEEE. https://doi.org/DOI: 10.1109/ICAC2N63387.2024.10894873

Madan, K., Gautam, D., Bhatia, S., & Kashyap, S. Chanchal, & Priyanka. (2023). LoRaWAN enabled wildland safety fire alert system. In *2023 14th International Conference on Computing Communication and Networking Technologies (ICCCNT)* (pp. 1–7). IEEE. https://doi.org/DOI: 10.1109/ICCCNT56998.2023.10306956

Nasirahmadi, A., & Hensel, O. (2022). Toward the next generation of digitalization in agriculture based on digital twin paradigm. *Sensors (Basel)*, *22*(2), 498. PMID: 35062459

Navarro, E., Costa, N., & Pereira, A. (2020). A systematic review of IoT solutions for smart farming. *Sensors (Basel)*, *20*, 4231. DOI: 10.3390/s20154231 PMID: 32751366

O'Flynn, B., Martinez-Catala, R., Harte, S., O'Mathuna, C., Cleary, J., Slater, C., & Murphy, H. (2007). "SmartCoast: A Wireless Sensor Network for Water Quality Monitoring, *32nd IEEE Conference on Local Computer Networks (LCN 2007)*. doi:DOI: 10.1109/lcn.2007.34

Oladunmoye, M., Oluwatomi, A. A., & Obakin, O. (2014). Design and construction of an automatic sliding door using infrared sensor. Computing, Information Systems, Development Informatics & Allied Research Journal, 5(4).

Purcell, W., Neubauer, T., & Mallinger, K. (2023). Digital Twins in agriculture: Challenges and opportunities for environmental sustainability. *Current Opinion in Environmental Sustainability*, *61*, 101252.

Quy, V. K., Hau, N. V., Anh, D. V., Quy, N. M., Ban, N. T., Lanza, S., Randazzo, G., & Muzirafuti, A. (•••). IoT-Enabled Smart Agriculture: Architecture, Applications, and Challenges. *Applied Sciences (Basel, Switzerland)*, *12*, 3396. DOI: 10.3390/ app12073396

Rajak, P., Ganguly, A., Adhikary, S., & Bhattacharya, S. (2023, December). Internet of Things and smart sensors in agriculture: Scopes and challenges. *Journal of Agriculture and Food Research*, *14*. Advance online publication. DOI: 10.1016/j.jafr.2023.100776

Ranjitha, R., Darshana Jain, N., Deepika, J., Deepika, S., & Hima, M. (2022), "A Novel IoT based Smart Garbage Monitoring System", *6th International Conference on Intelligent Computing and Control Systems (ICICCS)*, 2022.

Shahab, H., Iqbal, M., Sohaib, A., Khan, F. U., & Waqas, M. (2024, May). IoT-based agriculture management techniques for sustainable farming: A comprehensive review. *Computers and Electronics in Agriculture, 220.* Advance online publication. DOI: 10.1016/j.compag.2024.108851

Singh, K., Yadav, M., Yadav, R. K., Naib, B. B., Bhatia, S., & Panda, D. K. (2024, December). Prognosis of Relocation Disease in Animals using Aggregation Method with Optimization Techniques. In *2024 Eighth International Conference on Parallel, Distributed and Grid Computing (PDGC)* (pp. 15-20). IEEE.

Srinivasan, R. (2023). Python and OpenCV for sign language recognition. In *2023 International Conference on Device Intelligence, Computing and Communication Technologies (DICCT)* (pp. 1–5). IEEE. https://doi.org/DOI: 10.1109/DICCT56244.2023.10110225

Sushanth, G., & Sujatha, S. (2018), "IOT Based Smart Agriculture System", *2018 International Conference on Wireless Communications, Signal Processing and Networking (WiSPNET).* doi:DOI: 10.1109/wispnet.2018.8538702

Vaish, R., Al-Shaikh, S. M., Shah, S. K., Dhabliya, D., & Devi, R. V. (2024). Automated smart crop protection utilising IoT: a comprehensive review. In *E3S Web of Conferences* (Vol. 540, p. 13012). EDP Sciences.

Valarmathi, G., Suganthi, S. U., Subashini, V., Janaki, R., Sivasankari, R., & Dhanasekar, S. (2021). CNN algorithm for plant classification in deep learning. *Materials Today: Proceedings, 46,* 3684–3689. DOI: 10.1016/j.matpr.2021.01.847

Verdouw, C., Tekinerdogan, B., Beulens, A., & Wolfert, S. (2021). Digital twins in smart farming. *Agricultural Systems, 189,* 103046.

Vitali, G., Francia, M., Golfarelli, M., & Canavari, M. (2023). Crop Management with the IoT: An Interdisciplinary Survey. *Agronomy (Basel), 11,* 181. DOI: 10.3390/agronomy11010181

Wang, S. W., Chen, C. C., Wu, C. M., & Huang, C. M. (2018, April). A continuous water-level sensor based on load cell and floating pipe. In *2018 IEEE International Conference on Applied System Invention (ICASI)* (pp. 151-154). IEEE.

Xuan, Z., Fu, X., & Li, D. (2012, June). Monitoring microsystem in small shelter system used for ammunition transportation. In World Automation Congress 2012 (pp. 1-3). IEEE.

Compilation of References

Aase, J., & Schaefer, G. (1996). Economics of Tillage Practices and Spring Wheat and Barley Crop Sequences in the Northern Great Plains. *Journal of Soil and Water Conservation, 15*, 167–170.

Aggarwal, C. C. (2018). *Neural Networks and Deep Learning: A Textbook*. Springer International Publishing.

Ahmed, G. H. (2023). A review of '3D concrete printing': Materials and process characterization, economic considerations and environmental sustainability. *Journal of Building Engineering, 66*, 105863. DOI: 10.1016/j.jobe.2023.105863

Ahmed, G. H. (2023). A review of "3D concrete printing": Materials and process characterization, economic considerations and environmental sustainability. *Journal of Building Engineering, 66*, 105863.

Ahmed, L. J., Dhanasekar, S., Govindaraj, V., & Ezhilazhagan, C. (2023). Crop Management System. *Advanced Technologies for Smart Agriculture*, (Dec), 55–79. DOI: 10.1201/9781032628745-3

Angin, P., Anisi, M. H., Göksel, F., Gürsoy, C., & Büyükgülcü, A. (2020). AgriLoRa: A digital twin framework for smart agriculture. *Journal of Wireless Mobile Networks, Ubiquitous Computing and Dependable Applications, 11*(4), 77–96.

Ariffin, M. A. M., Ramli, M. I., Amin, M. N. M., Ismail, M., Zainol, Z., Ahmad, N. D., & Jamil, N. (2020). ''Automatic Climate Control for Mushroom Cultivation Using IoT Approach,'' *In Proceedings of the 2020 IEEE 10th International Conference on System Engineering and Technology (ICSET)*, Shah Alam, Malaysia, pp. 123–128.

Arz Von Straussenburg, A. F., Blazevic, M., & Riehle, D. M. (2023). Measuring the actual office workspace utilization in a desk sharing environment based on IoT sensors. In *Proceedings of the 18th International Conference on Design Science Research in Information Systems and Technology (DESRIST 2023): Design Science Research for a New Society: Society 5.0* (pp. 69–83). Lecture Notes in Computer Science. Pretoria, South Africa, May 31–June 2.

ASTM Committee on Additive Manufacturing Technologies ACF, on Additive Manufacturing Technologies. Subcommittee F42. 91 on Terminology ACF (2012) Standard terminology for additive manufacturing technologies. *ASTM International.*

Attaran, S., & Attaran, M. (2024). Ce lik BG. Digital Twins and Industrial Internet of Things: Uncovering operational intelligence in industry 4.0. *Decision Analytics Journal, 10*, 100398. DOI: 10.1016/j.dajour.2024.100398

Balasubramaniyan, M., & Navaneethan, C. (2021), "Applications of Internet of Things for smart farming – A survey", *Materials today proceedings, 12th National Conference on Recent Advancements in Biomedical Engineering*, Vol. 47, Part 1, 2021, Pages 18-24, https://doi.org/DOI: 10.1016/j.matpr.2021.03.480

Balathandapani, R., Boopathi, D., Jotheeshwaran, S., Arundeva, G., & Saranya, C. (2015). Automatic rain water and crop saving system using embedded technology. International Journal of Science, Engineering and Technology Research (IJSETR), 4(3).

Barn, B. S. (2022). The Sociotechnical Digital Twin: On the Gap between Social and Technical Feasibility. *In Proceedings of the 24th IEEE International Conference on Business Informatics,* Amsterdam, The Netherlands, 15–17 June 2022; pp. 11–20.

Belhajem, I., Ben Maissa, Y., & Tamtaoui, A. (2016). A Robust Low Cost Approach for Real Time Car Positioning in a Smart City Using Extended Kalman Filter and Evolutionary Machine Learning. *In Proceedings of the 2016 4th IEEE International Colloquium on Information Science and Technology (CiSt),* Tangier, Morocco, 24–26 October 2016; IEEE: Piscataway, NJ, USA, 2016; pp. 806–811.

Bhatia, S., Bhadauria, R. V. S., Mehrotra, K., Malik, J., Goel, A. K., & Naib, B. B. (2023). A comparative analysis and design criteria of MAC, SMAC, and TMAC protocols in wireless sensor network. In *2023 14th International Conference on Computing Communication and Networking Technologies (ICCCNT)* (pp. 1–6). IEEE. https://doi.org/DOI: 10.1109/ICCCNT56998.2023.10306758

Bhatia, S., Gautam, D., Kumar, S., & Verma, S. (2023). Automatic seat identification system in smart transport using IoT and image processing. In *2023 3rd International Conference on Intelligent Communication and Computational Techniques (ICCT)* (pp. 1–6). IEEE. https://doi.org/DOI: 10.1109/ICCT56969.2023.10075664

Bhatia, S., Jaffery, Z. A., & Mehfuz, S. (2023). A comparative study of wireless communication protocols for use in smart farming framework development. In *2023 3rd International Conference on Intelligent Communication and Computational Techniques (ICCT)* (pp. 1–7). IEEE. https://doi.org/DOI: 10.1109/ICCT56969.2023.10075696

Bhatia, S., Mallikarjun, B., Naib, B. B., & Goel, A. K. (2023). Algo-indicator based trading in stock market. In *2023 4th International Conference on Smart Electronics and Communication (ICOSEC)* (pp. 1730–1734). IEEE. https://doi.org/DOI: 10.1109/ICOSEC58147.2023.10275858

Bhatia, S., Naib, B. B., Goel, A. K., Kumar, L., & Singh, K. (2023). Drowsy driver alert system using OpenCV. In *2023 3rd International Conference on Pervasive Computing and Social Networking (ICPCSN)* (pp. 614–619). IEEE. https://doi.org/DOI: 10.1109/ICPCSN58827.2023.00108

Bhatia, S., Arya, C., Verma, S., Gautam, D., Naib, B. B., & Kumar, A. (2023). Predict success of a Zomato restaurant using machine learning. In *2023 World Conference on Communication & Computing (WCONF)* (pp. 1–7). IEEE. https://doi.org/DOI: 10.1109/WCONF58270.2023.10235064

Bhatia, S., Dhillon, H. S., & Kumar, N. (2011). Alive human body detection system using an autonomous mobile rescue robot. In *2011 Annual IEEE India Conference* (pp. 1–5). IEEE. https://doi.org/DOI: 10.1109/INDCON.2011.6139388

Bhatia, S., Goel, A. K., Naib, B. B., Singh, K., Yadav, M., & Saini, A. (2023). Diabetes prediction using machine learning. In *2023 World Conference on Communication & Computing (WCONF)* (pp. 1–6). IEEE. https://doi.org/DOI: 10.1109/WCONF58270.2023.10235187

Bhatia, S., Goel, N., Ahlawat, V., Naib, B. B., & Singh, K. (2023). A comprehensive review of IoT reliability and its measures: Perspective analysis. In *Handbook of research on machine learning-enabled IoT for smart applications across industries* (pp. 365–384). IGI Global.

Bhatia, S., Goel, N., Dandeva, D., Sharma, N., Mallikarjun, B., & Naib, B. B. (2024). Detection of posture using MS-IoT-based solution to provide timely treatment. In *Using machine learning to detect emotions and predict human psychology* (pp. 209–234). IGI Global.

Bhatia, S., Jaffery, Z. A., & Mehfuz, S. (2023). Development and analysis of IoT based smart agriculture system for heterogenous nodes. In *2023 International Conference on Recent Advances in Electrical, Electronics & Digital Healthcare Technologies (REEDCON)* (pp. 62–67). IEEE. https://doi.org/DOI: 10.1109/REEDCON57544.2023.10150466

Bhatia, S., Mallikarjuna, B., Gautam, D., Gupta, U., Kumar, S., & Verma, S. (2023). The future IoT: The current generation 5G and next generation 6G and 7G technologies. In *2023 International Conference on Device Intelligence, Computing and Communication Technologies (DICCT)* (pp. 212–217). IEEE. https://doi.org/DOI: 10.1109/DICCT56244.2023.10110066

Bhatia, S., Naib, B. B., Goel, N., Agarwal, B., Mallikarjun, B., & Arvindhan, M. (2024). A new era in weather forecasting: Harnessing the potential of machine learning. In *2024 IEEE International Conference on Computing, Power and Communication Technologies (IC2PCT)* (pp. 1355–1361). IEEE. https://doi.org/DOI: 10.1109/IC2PCT60090.2024.10486245

Bosch, D., & Napit, K. (1992). Economics of Transferring Poultry Litter to Achieve More Effective Use as Fertilizer. *Journal of Soil and Water Conservation*, *47*(July-August).

Boyanov, L., Kisimov, V., & Christov, Y. (2020). Evaluating IoT Reference Architecture. *In Proceedings of the 2020 International Conference Automatics and Informatics (ICAI),* Varna, Bulgaria, pp. 1–5.

Bultena, G., & Hoiberg, E. (1983). Factors Affecting Farmers' Adoption of Conservation Tillage. *Journal of Soil and Water Conservation*, (May-June), 281–284.

Camille, L. T., Sullivan, J. J., Drapala, K. P., Chartrin, V., Chan, T., Morrison, A. P., & Kerry, J. P., & 620 Kelly, A. L. (2017). Effect of 3D printing on the structure and textural properties of processed 621 cheese. *Journal of Food Engineering*.

Carmody, J., Shringarpure, S., & Van de Venter, G. (2021). AI and privacy concerns: A smart meter case study. *Journal of Information. Communication and Ethics in Society*, *19*(4), 492–505.

Castro, S., Iñacasha, J., Mesias, G., & Oñate, W. (2023). Prototype Based on a LoRaWAN Network for Storing Multivariable Data, Oriented to Agriculture with Limited Resources. *In Proceedings of the Seventh International Congress on Information and Communication Technology, London,* UK, 21–24 February 2022; Springer Nature: Singapore, pp. 245–255.

Cesco, S., Sambo, P., Borin, M., Basso, B., Orzes, G., & Mazzetto, F. (2023). Smart agriculture and digital twins: Applications and challenges in a vision of sustainability. *European Journal of Agronomy*, *146*, 126809.

Cesco, S., Zara, V., De Toni, A. F., Lugli, P., Betta, G., Evans, A. C., & Orzes, G. (2021). Higher Education in the First Year of COVID-19: Thoughts and Perspectives for the Future. *International Journal of Higher Education, 10*, •••. https://www.sciedu.ca/journal/index.php/ijhe/article/view/20207

Chan, C. K. Y., & Hu, W. (2023). Students' voices on generative AI: Perceptions, benefits, and challenges in higher education. *International Journal of Educational Technology in Higher Education*, *20*(1), 43.

Chen, A., & Drake, O. (2023, August 8). Chen, *Accuracy of Chatbots in Citing Journal Articles. JAMA Network Open*. Advance online publication. DOI: 10.1001/jamanetworkopen.2023.27647 PMID: 37552482

Chen, C., Fu, H., Zheng, Y., Tao, F., & Liu, Y. (2023). The advance of digital twin for predictive maintenance: The role and function of machine learning. *Journal of Manufacturing Systems*, *71*, 581–594. DOI: 10.1016/j.jmsy.2023.10.010

Concepcion, R., Josh Ramirez, T., Alejandrino, J., Janairo, A. G., Jahara Baun, J., Francisco, K., Relano, R.-J., Enriquez, M. L., Grace Bautista, M., Vicerra, R. R., ... (2022). A Look at the Near Future: Industry 5.0 Boosts the Potential of Sustainable Space Agriculture. *In Proceedings of the 2022 IEEE 14th International Conference on Humanoid, Nanotechnology, Information Technology, Communication and Control, Environment, and Management (HNICEM),* Boracay Island, Philippines, 1–4 December 2022; IEEE: Piscataway, NJ, USA, pp. 1–6.

De Alwis, S., Hou, Z., Zhang, Y., & Na, M. H. (2022, June). "A survey on smart farming data, applications and techniques. *Computers in Industry*, *138*. Advance online publication. DOI: 10.1016/j.compind.2022.103624

Derossi, A., Caporizzi, R., Azzollini, D., & Severini, C. (2018). Application of 3D printing for customized food. A case on the development of a fruit-based snack for children. *Journal of Food Engineering*, *220*, 65–75.

Dhillon, H. S., Chawla, P., Kumar, K., & Bhatia, S. (2024). Study the effect of hybrid energy harvesting wireless body sensor nodes using multi-dimensional approach in simulation. In *2024 IEEE 5th India Council International Subsections Conference (INDISCON)* (pp. 1–5). IEEE. https://doi.org/DOI: 10.1109/INDISCON62179.2024.10744340

Dizon, J. R. C., Valino, A. D., Souza, L. R., Espera, A. H., Jr., Chen, Q., & Advincula, R. C. "3Dprinted molds and materials for injection molding and rapid tooling applications," *MRS Communications Prospectives Journal,* vol. 9, issue 4. 2019.

Dovidio, J. F., Kawakami, K., & Gaertner, S. L. (2002). Implicit and explicit prejudice and interracial interaction. *Journal of Personality and Social Psychology, 82*(1), 62. PMID: 11811635

Edemetti, F., Maiale, A., Carlini, C., D'Auria, O., Llorca, J., & Maria Tulino, A. (2022). Vineyard Digital Twin: construction and characterization via UAV images—DIWINE Proof of Concept. *In Proceedings of the 2022 IEEE 23rd International Symposium on a World of Wireless,* Mobile and Multimedia Networks (WoWMoM), Belfast, UK, pp. 601–606.

Edemetti, F., Maiale, A., Carlini, C., D'Auria, O., Llorca, J., & Tulino, A. M. (2022). Vineyard Digital Twin: Construction and characterization via UAV images–DIWINE Proof of Concept. *In Proceedings of the 2022 IEEE 23rd International Symposium on a World of Wireless, Mobile and Multimedia Networks (WoWMoM),* Belfast, UK, pp. 601–606.

Elahi, A. F., & Rahman, M. S. (2014, December). Intelligent windshield for automotive vehicles. In 2014 17th international conference on computer and information technology (ICCIT) (pp. 392-396). IEEE.

Fraser, A. (2022). 'You can't eat data'?: Moving beyond the misconfigured innovations of smart farming. *Journal of Rural Studies, 91*, 200–207. DOI: 10.1016/j.jrurstud.2021.06.010

Fred, L. A., Lohrengel, A., Neubert, V., Camila, F. H., & Czelusniak, T. (2014). Selective laser 638 sintering of Mo-CuNi composite to be used as EDM electrode. *Rapid Prototyping Journal,* 20, 639 59-68.

Fuentealba, D., Flores, C., Soto, I., Zamorano, R., & Reid, S. (2022). Guidelines for digital twins in 5g agriculture. *In Proceedings of the 2022 13th International Symposium on Communication Systems, Networks and Digital Signal Processing (CSNDSP),* Porto, Portugal, pp. 613–618.

Fuentealba, D., Flores, C., Soto, I., Zamorano, R., & Reid, S. (2022). Guidelines for Digital Twins in 5G Agriculture. *In Proceedings of the 2022 13th International Symposium on Communication Systems, Networks and Digital Signal Processing (CSNDSP),* Porto, Portugal, pp. 613–618.

Ganai, P. T., Bag, A., Sable, A., Abdullah, K. H., Bhatia, S., & Pant, B. (2022). A detailed investigation of implementation of Internet of Things (IoT) in cyber security in healthcare sector. In *2022 2nd International Conference on Advance Computing and Innovative Technologies in Engineering (ICACITE)* (pp. 1571–1575). IEEE. https://doi.org/DOI: 10.1109/ICACITE53722.2022.9823887

Gardezi, M., Adereti, D. T., Stock, R., & Ogunyiola, A. (2022). In pursuit of responsible innovation for precision agriculture technologies. *Journal of Responsible Innovation*, *9*(2), 224–247. DOI: 10.1080/23299460.2022.2071668

Gautam, D., Bhatia, S., Goel, N., & Mallikaijuna, B., H S, G., & Naib, B. B. (2023). Development of IoT enabled framework for LPG gas leakage detection and weight monitoring system. In *2023 International Conference on Device Intelligence, Computing and Communication Technologies (DICCT)* (pp. 182–187). IEEE. https://doi.org/DOI: 10.1109/DICCT56244.2023.10110294

Gautam, D., Bhatia, S., Neha, P., & Naib, B. B. (2023). IoT based automatic traffic light and speed breaker system. In *2023 3rd International Conference on Innovative Sustainable Computational Technologies (CISCT)* (pp. 1–6). IEEE. https://doi.org/DOI: 10.1109/CISCT57197.2023.10351445

Goel, N., Bhatia, S., & Rana, P. (2023). Secure smart devices control with low power consumption. In *2023 3rd International Conference on Advancement in Electronics & Communication Engineering (AECE)* (pp. 83–87). IEEE.

Gogul Dev, N. S., Sreenesh, K. S., & Binu, P. K. " IoT Based Automated Crop Protection System," *2019 2nd International Conference on Intelligent Computing, Instrumentation and Control Technologies (ICICICT)*. doi:, 2019.DOI: 10.1109/icicict46008.2019.899

Govindaraj, V., Dhanasekar, S., Martinsagayam, K., Pandey, D., Pandey, B. K., & Nassa, V. K. (2023, January). Low-power test pattern generator using modified LFSR. *Aerospace Systems*, *7*(1), 67–74. DOI: 10.1007/s42401-022-00191-5

Greenstein, M. N. (2020). AI and a Judge's Ethical Obligations. *The Judges' Journal*, *59*, 40.

Grieves, M. W. (2012). Virtually Indistinguishable. *In Proceedings of the PLM: IFIP International Conference on Product Lifecycle Management, Montreal, QC, Canada.*

Hameed, I. A. (2018). A Coverage Planner for Multi-Robot Systems in Agriculture. *In Proceedings of the IEEE International Conference on Real-time Computing and Robotics (RCAR),* Kandima, Maldives, pp. 698–704.

Hamilton, C. A., Alici, G., & Marc, P. (2017). 3D printing Vegemite and Marmite: Redefining 649 "breadboards". Journal of Food Engineering.

Harris, E., Khoo, I. H., Yi, C., & Demircan, E. (2024). A protocol for evaluating the performance of self-powered sensing units in human activity recognition using triboelectric nanogenerators. In *Proceedings of the International Conference on Advanced Robotics and Mechatronics* (pp. 120–125). Tokyo, Japan, July 8–10.

Hemsley, B., Palmer, S., Kouzani, A., Adams, S., & Balandin, S. (2019). Review informing the design of 3D food printing for people with swallowing disorders: Constructive, conceptual, and empirical problems. In *Proceedings of the Hawaii International Conference on System Sciences*. https://doi.org/DOI: 10.24251/HICSS.2019.692

Howard, D. A., Ma, Z., Aaslyng, J. M., & Jørgensen, B. N. Data Architecture for Digital Twin of Commercial Greenhouse Production. *In Proceedings of the 2020 RIVF International Conference on Computing and Communication Technologies (RIVF),* Ho Chi Minh City, Vietnam, 14–15 October 2020; pp. 1–7.

Hunt, E. R., & Daughtry, C. S. T. (2018). What good are unmanned aircraft systems for agricultural remote sensing and precision agriculture? *International Journal of Remote Sensing*, *39*(21), 5345–5376.

Hurst, W., Spyrou, O., & Krampe, C. (2024). The Data Factory: Findings from an Extended Reality-Based Hackathon for Data Science Education. *In Proceedings of the 2024 10th International Conference on Virtual Reality (ICVR),* Bournemouth, UK, pp. 85–90.

International Society of Sports Nutrition (ISSN). (2018). Proceedings of the Fifteenth International Society of Sports Nutrition (ISSN) Conference and Expo: Clearwater Beach, FL, USA, June 7–9, 2018. *Journal of the International Society of Sports Nutrition, 15*(S1), 1–37. DOI: 10.1186/s12970-018-0256-5

Jackson, R. (1984). Remote Sensing of Vegetation Characteristics for Farm Management. *Proceedings of SPIE - the International Society for Optical Engineering, 475*, 81–96.

Jain, A., Roy, D. K., Ayasrah, F. T., William, P., & Kulkarni, S. (2024). "Enhancing Metaverse Cybersecurity With Federated Learning," *Global Congress on Emerging Technologies (GCET-2024)*, Gran Canaria, Spain, pp. 254-265, doi: DOI: 10.1109/GCET64327.2024.10934363

Javaid, M., Haleem, A., Singh, R. P., & Suman, R. (2022, January). Enhancing smart farming through the applications of Agriculture 4.0 technologies. *International Journal of Intelligent Networks*, *3*, 150–164.

Jeet, V., Dhillon, H. S., & Bhatia, S. (2015). Radio frequency home appliance control based on head tracking and voice control for disabled person. In *2015 Fifth International Conference on Communication Systems and Network Technologies* (pp. 559–563). IEEE. https://doi.org/DOI: 10.1109/CSNT.2015.189

Jiao, X., Chai, S., & Wang, S. (2010). Virtual assembly of agriculture equipment key components based on virtual reality technique, *in: 2010 Third International Conference on Information and Computing, IEEE,* Wuxi, TBD, China, pp. 27–30, https://doi.org/DOI: 10.1109/ICIC.2010.13

John, A. M., Aiswarya, M., & Panachakel, J. T. "Ethical challenges of using artificial intelligence in judiciary," *in 2023 IEEE International Conference on Metrology for eXtended Reality, Artificial Intelligence and Neural Engineering (MetroXRAINE).* IEEE, 2023, pp. 723–728.

Jones, D., Snider, C., Nassehi, A., Yon, J., & Hicks, B. (2020). *Characterising the digital twin: A systematic literature review* (Vol. 29). CIRP Journal of Manufacturing Science and Technology. [Online], Available https://www.sciencedirect.com/science/article/pii/S1755581720300110

Jo, S. K., Park, D. H., Park, H., & Kim, S. H. (2018). Smart livestock farms using digital twin: Feasibility study. In *Proceedings of the 2018 International Conference on Information and Communication Technology Convergence (ICTC)*, Jeju Island, Republic of Korea, 17–19 October 2018; pp. 1461–1463.

Jo, S. K., Park, D. H., Park, H., & Kim, S. H. (2018). Smart livestock farms using Digital Twin: feasibility study. *In: 2018 International Conference on Information and Communication Technology Convergence (ICTC),* pp. 1461–1463. https://doi.org/DOI: 10.1109/ICTC.2018.8539516

Kalaiselvi, K., Anand, A. J., Tanwar, P., & Raza, H. (Eds.). (2024). *Advanced technologies for smart agriculture*. CRC Press.

Kalogiannidis, S., Kalfas, D., Chatzitheodoridis, F., & Papaevangelou, O. (2024). Role of Crop-Protection Technologies in Sustainable Agricultural Productivity and Management. *Land (Basel)*, *11*. Advance online publication. DOI: 10.3390/land11101680

Kalyani, Y., Bermeo, N. V., & Collier, R. (2023). Digital twin deployment for smart agriculture in Cloud-Fog-Edge infrastructure. *. *International Journal of Parallel, Emergent and Distributed Systems : IJPEDS*, *38*(6), 1–16. DOI: 10.1080/17445760.2023.2235653

Kulkarni, S. A., Raikar, V. D., Rahul, B. K., Rakshitha, L. V., Sharanya, K., & Jha, V. (2020). '' Intelligent Water Level Monitoring System Using IoT,'' *IEEE International Symposium on Sustainable Energy, Signal Processing and Cyber Security (iSSSC)*. doi:DOI: 10.1109/isssc50941.2020.93588,2020

Kumar , V., & Garg, M. L. (2018). Predictive analytics: A review of trends and techniques. *International Journal of Computer Applications*, *182*, 31–37. DOI: 10.5120/ijca2018917434

Kumar, D., Bhatia, S., Dhillon, H. S., & Goel, A. K. (2024). NLP-based sentiment analysis using deep learning methods. In *2024 1st International Conference on Advances in Computing, Communication and Networking (ICAC2N)* (pp. 1873–1878). IEEE. https://doi.org/DOI: 10.1109/ICAC2N63387.2024.10894873

Lavanaya, M., & Parameswari, R. (2018). Soil nutrients monitoring for greenhouse yield enhancement using pH value with IoT and wireless sensor network. In *Proceedings of the 2018 Second International Conference on Green Computing and Internet of Things (ICGCIoT)* (pp. 547–552). IEEE.

Lepenioti, K.. (2020). Prescriptive analytics: *Literature review and research challenges. International Journal of Information Management, 50*, 57–70. DOI: 10.1016/j.ijinfomgt.2019.04.003

Lin, N., Wang, X., Zhang, Y., Hu, X., & Ruan, J. (2020). Fertigation management for sustainable precision agriculture based on Internet of Things. *Journal of Cleaner Production, 277*. Advance online publication. DOI: 10.1016/j.jclepro.2020.124119

Madan, K., Gautam, D., Bhatia, S., & Kashyap, S. Chanchal, & Priyanka. (2023). LoRaWAN enabled wildland safety fire alert system. In *2023 14th International Conference on Computing Communication and Networking Technologies (ICCCNT)* (pp. 1–7). IEEE. https://doi.org/DOI: 10.1109/ICCCNT56998.2023.10306956

Montavon, G., Samek, W., & Müller, K. R. (2018). Methods for interpreting and understanding deep neural networks. Digital Signal Processing: A Review Journal, 73, 1–15. doi: .DOI: 10.1016/j.dsp.2017.10.011

Monteiro, J., Barata, J., Veloso, M., Veloso, L., & Nunes, J. (2018). Towards Sustainable Digital Twins for Vertical Farming. *In Pro ceedings of the 2018 Thirteenth International Conference on Digital Information Management (ICDIM),* Berlin, Germany, pp. 234–239.

Monteiro, J., Barata, J., Veloso, M., Veloso, L., & Nunes, J. (2018). Towards Sustainable Digital Twins for Vertical Farming. *In Proceedings of the 2018 Thirteenth International Conference on Digital Information Management (ICDIM),* Berlin, Germany, pp. 234–239.

Monteiro, J., Barata, J., Veloso, M., Veloso, L., & Nunes, J. (2018). Towards sustainable digital twins for vertical farming. *In Proceedings of the 2018 Thirteenth International Conference on Digital Information Management (ICDIM),* Berlin, Germany; pp. 234–239.

Mulyani, G. S., Adhitya, Y., & Köppen, M. (2023). Design and Implementation of Farmer Digital Twin Control in Smart Farming. *In Proceedings of the International Conference on Intelligent Networking and Collaborative Systems,* Chiang Mai, Thailand, 6–8 September 2023; Springer: Berlin/Heidelberg, Germany, pp. 516–527.

Nasirahmadi, A., & Hensel, O. (2022). Toward the next generation of digitalization in agriculture based on digital twin paradigm. *Sensors (Basel)*, *22*(2), 498. PMID: 35062459

Navarro, E., Costa, N., & Pereira, A. (2020). A systematic review of IoT solutions for smart farming. *Sensors (Basel)*, *20*, 4231. DOI: 10.3390/s20154231 PMID: 32751366

Nie, J., Wang, Y., Li, Y., & Chao, X. N. I. E. (2022). Artificial intelligence and digital twins in sustainable agriculture and forestry: A survey. *Turkish Journal of Agriculture and Forestry*, *46*(5), 5. DOI: 10.55730/1300-011X.3033

Nigam, A., Kabra, P., & Doke, P. (2011). Augmented reality in agriculture, *in: 2011 IEEE 7th International Conference on Wireless and Mobile Computing, Networking and Communications (WiMob),* IEEE, Shanghai, China, pp. 445–448, https://doi.org/DOI: 10.1109/WiMOB.2011.6085361

Nyman, H. J., & Sarlin, P. (2014) From bits to atoms: 3D printing in the context of supply chain strategies. *In: 2014 47th Hawaii international conference on system sciences.* Pp 4190–4199

O'Flynn, B., Martinez-Catala, R., Harte, S., O'Mathuna, C., Cleary, J., Slater, C., & Murphy, H. (2007). "SmartCoast: A Wireless Sensor Network for Water Quality Monitoring, *32nd IEEE Conference on Local Computer Networks (LCN 2007).* doi:DOI: 10.1109/lcn.2007.34

Oladunmoye, M., Oluwatomi, A. A., & Obakin, O. (2014). Design and construction of an automatic sliding door using infrared sensor. Computing, Information Systems, Development Informatics & Allied Research Journal, 5(4).

Otieno, M. (2023). An extensive survey of smart agriculture technologies: Current security posture. *World Journal of Advanced Research and Reviews, 18*(3), 1207–1231. DOI: 10.30574/wjarr.2023.18.3.1241

Padhiary, M., Rani, N., Saha, D., Barbhuiya, J. A., & Sethi, L. N. (2023). Efficient Precision Agriculture with Python-based Raspberry Pi Image Processing for Real-Time Plant Target Identification. *IJRAR, 10*(3), 539–545. DOI: 10.1729/Journal.35531

Panachakel, J. T., Sharma, K., Anusha, A., & Ramakrishnan, A. (2021). "Can we identify the category of imagined phoneme from EEG?" *in 2021 43rd Annual International Conference of the IEEE Engineering in Medicine & Biology Society (EMBC).* IEEE, pp. 459–462.

Panarin, R. N., & Khvorova, L. A. (2022). Software Development for Agricultural Tillage Robot Based on Technologies of Machine Intelligence. *In Proceedings of the International Conference on High-Performance Computing Systems and Technologies in Scientific Research, Automation of Control and Production,* Barnaul, Russia, 20–21 May 2022; Springer: Cham, Switzerland, pp. 354–367.

Podchasoy, E. O. (2021). Designd and Technological Features on 3D-Printing Usage in Agricultural Machines Gearings Repair. *International Journal of Mechanical Engineering and Robotics Research, 10*(1), 32–37.

Purcell, W., Neubauer, T., & Mallinger, K. (2023). Digital Twins in agriculture: Challenges and opportunities for environmental sustainability. *Current Opinion in Environmental Sustainability, 61*, 101252.

Quy, V. K., Hau, N. V., Anh, D. V., Quy, N. M., Ban, N. T., Lanza, S., Randazzo, G., & Muzirafuti, A. (•••). IoT-Enabled Smart Agriculture: Architecture, Applications, and Challenges. *Applied Sciences (Basel, Switzerland), 12*, 3396. DOI: 10.3390/ app12073396

Rajak, P., Ganguly, A., Adhikary, S., & Bhattacharya, S. (2023, December). Internet of Things and smart sensors in agriculture: Scopes and challenges. *Journal of Agriculture and Food Research, 14*. Advance online publication. DOI: 10.1016/j.jafr.2023.100776

Ranjitha, R., Darshana Jain, N., Deepika, J., Deepika, S., & Hima, M. (2022), "A Novel IoT based Smart Garbage Monitoring System*", 6th International Conference on Intelligent Computing and Control Systems (ICICCS)*, 2022.

Rawankar, A., ... (2018). "Detection of N, P, K fertilizers in agricultural soil with NIR laser absorption technique," *in 2018 3rd International Conference on Microwave and Photonics (ICMAP),* Feb. 2018, pp. 1–2. doi: DOI: 10.1109/ICMAP.2018.8354625

Rehman, H. U., Asif, M., & Ahmad, M. (2017). Future applications and research challenges of IoT. In *Proceedings of the International Conference on Information and Communication Technologies* (pp. 68–74). Karachi, Pakistan, December 30–31.

Rodrigues, R. (2020). Legal and human rights issues of AI: Gaps, challenges and vulnerabilities. *Journal of Responsible Technology, 4*, 100005.

Rojko, A. (2017). Industry 4.0 Concept. *Background and Overview International Journal of Interactive Mobile Technologies, 11*, 77–90.

Rudd, J. D., Roberson, G. T., & Classen, J. J. (2017). Application of satellite, unmanned aircraft system, and ground-based sensor data for precision agriculture: A review. *In Proceedings of the 2017 ASABE Annual International Meeting;* American Society of Agricultural and Biological Engineers, Spokane, WA, USA.

Sarma, R., & Barbhuiya, F. A. (2019). Internet of Things: Attacks and defences. In *Proceedings of the 2019 7th International Conference on Smart Computing & Communications (ICSCC)* (pp. 1–5). IEEE.

Schuh, J. C. L., & Funk, K. A. (2019, April). *—Compilation of International Standards and Regulatory Guidance Documents for Evaluation of Biomaterials, Medical Devices, and 3-D Printed and Regenerative Medicine Products,* ||. *Toxicologic Pathology, 47*(3), 344–357. DOI: 10.1177/0192623318804121 PMID: 30392453

Shahab, H., Iqbal, M., Sohaib, A., Khan, F. U., & Waqas, M. (2024, May). IoT-based agriculture management techniques for sustainable farming: A comprehensive review. *Computers and Electronics in Agriculture, 220*. Advance online publication. DOI: 10.1016/j.compag.2024.108851

Sharma, K., ... (2024). Framework for Image Processing-Based Robotic Car for Agricultural Ploughing Using Ensembling Machine Learning Approach. In: Tomar, A., Mishra, S., Sood, Y.R., Kumar, P. (eds) Proceedings of 4th International Conference on Machine Learning, Advances in Computing, Renewable Energy and Communication. MARC 2023. Lecture Notes in Electrical Engineering, vol 1231. Springer, Singapore. https://doi.org/DOI: 10.1007/978-981-97-5227-0_3

Shepherd, M., Turner, J. A., Small, B., & Wheeler, D. (2020). Priorities for science to overcome hurdles thwarting the full promise of the 'digital agriculture' revolution. *Journal of the Science of Food and Agriculture, 100*(14), 5083–5092. DOI: 10.1002/jsfa.9346 PMID: 30191570

Singh, K., Yadav, M., Yadav, R. K., Naib, B. B., Bhatia, S., & Panda, D. K. (2024, December). Prognosis of Relocation Disease in Animals using Aggregation Method with Optimization Techniques. In *2024 Eighth International Conference on Parallel, Distributed and Grid Computing (PDGC)* (pp. 15-20). IEEE.

Skobelev, P., Tabachinskiy, A., Simonova, E., Lee, T.-R., Zhilyaev, A., & Laryukhin, V. (2021). Digital twin of rice as a decision-making service for precise farming, based on environmental datasets from the fields. *In Proceedings of the 2021 International Conference on Information Technology and Nanotechnology (ITNT),* Samara, Russian Federation, pp. 1–8.

Srinivasan, R. (2023). Python and OpenCV for sign language recognition. In *2023 International Conference on Device Intelligence, Computing and Communication Technologies (DICCT)* (pp. 1–5). IEEE. https://doi.org/DOI: 10.1109/DICCT56244.2023.10110225

Stary, C., & Neubauer, M. (2017). Industrial Challenges. In Neubauer, M., & Stary, C. (Eds.), *S-BPM in the Production Industry* (pp. 7–25). Springer International Publishing.

Sushanth, G., & Sujatha, S. (2018), "IOT Based Smart Agriculture System", *2018 International Conference on Wireless Communications, Signal Processing and Networking (WiSPNET).* doi:DOI: 10.1109/wispnet.2018.8538702

Sushanth, G., & Sujatha, S. (2018). IOT based smart agriculture system. *In Proceedings of the 2018 International Conference on Wireless Communications, Signal Processing and Networking (WiSPNET),* Chennai, India, pp. 1–4.

Toupas, P., Tsamis, G., Zotou, D., Giakoumis, D., Votis, K., & Tzovaras, D. (2024). Harnessing IoT and robotics for comprehensive human activity recognition in smart buildings. In *Proceedings of the 5th Olympus International Conference (ICSC 2024)* (pp. 161–173). Katerini, Greece, May 24–26.

Vaish, R., Al-Shaikh, S. M., Shah, S. K., Dhabliya, D., & Devi, R. V. (2024). Automated smart crop protection utilising IoT: a comprehensive review. In *E3S Web of Conferences* (Vol. 540, p. 13012). EDP Sciences.

Vajpayee, P., & Hossain, G. (2024). "Multi-Tenant Cloud Security- Risk Prediction through Cyber-Value-at-Risk (CVaR)," *12th International Symposium on Digital Forensics and Security* (pp. 1-7), doi: DOI: 10.1109/ISDFS60797.2024.10527323

Valarmathi, G., Suganthi, S. U., Subashini, V., Janaki, R., Sivasankari, R., & Dhanasekar, S. (2021). CNN algorithm for plant classification in deep learning. *Materials Today: Proceedings*, *46*, 3684–3689. DOI: 10.1016/j.matpr.2021.01.847

Verdouw, C., Tekinerdogan, B., Beulens, A., & Wolfert, S. (2021). Digital twins in smart farming. *Agricultural Systems, 189*, 103046.

Vitali, G., Francia, M., Golfarelli, M., & Canavari, M. (2023). Crop Management with the IoT: An Interdisciplinary Survey. *Agronomy (Basel), 11*, 181. DOI: 10.3390/agronomy11010181

Wang, S. W., Chen, C. C., Wu, C. M., & Huang, C. M. (2018, April). A continuous water-level sensor based on load cell and floating pipe. In *2018 IEEE International Conference on Applied System Invention (ICASI)* (pp. 151-154). IEEE.

Wang, Z., Zhao, C., Zhang, H., & Fan, H. (2011). Real-time remote monitoring and warning system in general agriculture environment. In *Proceedings of the 2011 International Conference of Information Technology, Computer Engineering and Management Sciences* (pp. 160–163). IEEE.

William, P., Shrivastava, A., Aswal, U. S., Kumar, I., Gupta, M., & Rao, A. K. (2023). "Framework for Implementation of Android Automation Tool in Agro Business Sector," 2023 4th International Conference on Intelligent Engineering and Management (ICIEM), London, United Kingdom, pp. 1-6, doi: DOI: 10.1109/ICIEM59379.2023.10167328

William, P., Darwante, N. K., Pawar, A. B., Jawale, M. A., & Verma, A. (2023). Framework for Implementation of Smart Driver Assistance System Using Augmented Reality. In Marriwala, N., Tripathi, C., Jain, S., & Kumar, D. (Eds.), *Mobile Radio Communications and 5G Networks. Lecture Notes in Networks and Systems* (Vol. 588). Springer., DOI: 10.1007/978-981-19-7982-8_30

William, P., Jaiswal, V. K., Shrivastava, A., Bansal, S., Hussein, L., & Singla, A. (2025). "Digital Identity Protection: Safeguarding Personal Data in the Metaverse Learning," *2025 International Conference on Engineering, Technology & Management (ICETM)*, Oakdale, NY, USA, pp. 1-6, doi: DOI: 10.1109/ICETM63734.2025.11051435

Xuan, Z., Fu, X., & Li, D. (2012, June). Monitoring microsystem in small shelter system used for ammunition transportation. In World Automation Congress 2012 (pp. 1-3). IEEE.

Yogeesh, N., Girija, D. K., Rashmi, M., & William, P. (2024). Intelligent Irrigation Systems in Agriculture Using Fuzzy Logic Techniques. In Shukla, B., Murthy, B. K., Hasteer, N., Kaur, H., & Van Belle, J. P. (Eds.), *Intelligent IT Solutions for Sustainability in Industry 5.0 Paradigm. ICEIL 2023. Lecture Notes in Electrical Engineering* (Vol. 1185). Springer., DOI: 10.1007/978-981-97-1682-1_25

Zermas, D., Morellas, V., Mulla, D., & Papanikolopoulos, N. (2018). Extracting phenotypic characteristics of corn crops through the use of reconstructed 3D models. *In Proceedings of the 2018 IEEE/RSJ International Conference on Intelligent Robots and Systems (IROS),* Madrid, Spain, pp. 8247–8254.

About the Contributors

P. William is working as Director (Research) at Sanjivani University, Kopargaon, India. He is the Post Doctoral Fellow from Amity University Dubai, UAE and Adjunct faculty of Victorian Institute of Technology, Australia. He is recognized in World Top 2% Scientist list by Stanford University and Elsevier. He is a member of IEEE, QCFI, ISTE and various other professional bodies. His research includes innovation and development of cutting-edge solutions in the fields of natural language processing, artificial intelligence, deep learning, machine learning, soft computing, cybersecurity, and cloud computing. He has published 225+ papers in Scopus indexed journals and Conferences. He has 30+ patents published with grants in his credit. He has authored and edited 10+ books with renowned publishers of global recognition. He has been associated with numerous Multi-National Companies and various Educational Groups for his expertise in research, corporate training and consulting where he has contributed to the advancement of knowledge and practice in his domain. A focused professional with experience of consulting in Research, Innovation and Development. He served as a Chairperson and Auditor in multiple committees of national recognition. Delivered Keynote speeches and chaired many sessions in International Conferences. He was appointed as Series Editor, Guest Editor and Reviewer in Scopus/ Web of Science indexed journals.

Sharmila received B.E. degree in Electronics and Communication Engineering from the Annamalai University, the M.E degree in Computer and Communication engineering from Anna University, Tamil Nadu, and the Ph.D. degree in Electronics & Communication Engineering from Pondicherry University. She is an eminent academician having more than 14 years of teaching experience. Currently she is working with Raj Kumar Goel Institute of Technology, Ghaziabad as Assistant Professor in Electronics and Communication Engineering Department. She has published widely in International Journals and Conferences, her research findings related to Wireless Sensor Networks, Digital image Processing, Cryptography and

Information Security, and Block chain. She has authored/co-authored more than 40 research papers and 5 chapters in six other books, published by IET, Springer and Elsevier. She also contributes as an Editor of Books in CRC publication, & Bentham Science.

A. K. Haghi, PhD is a Professor Emeritus of Engineering Sciences and has published 250+ academic research-oriented books and 1000+ research papers in various journals and conference proceedings. His leadership in academic publishing includes founding and serving as Editor-in-Chief of prestigious journals. Professor Haghi's extensive educational background and supervisory roles underscore his expertise and contributions to the field of engineering sciences. He is currently appointed as Honorary Research Associate (HRA) at University of Coimbra, Portugal.

***

Sandeep Bhatia is a dedicated Associate Professor with more than 14 years of rich experience in teaching, research, and development activities. He is teaching undergraduate and post-graduate courses in the branch of CSE and ECE. He has one-year industry experience as a hardware engineer at C-DAC Mohali, Punjab. Presently, He is working as an Associate Professor in the Department of CSE, Galgotias University, Greater Noida. He also worked as Scientific officer (TIFAC-CORE) and Assistant Professor at AKGEC Ghaziabad. He is Pursuing Ph.D. from Jamia Millia Islamia, Central University, New Delhi in the field of IoT and Embedded System Design. He has completed his M.E from PEC University of Technology, Chandigarh in the field of Embedded System in the year 2010. He has completed his B.Tech from PTU Jalandhar in the year 2008. He has successfully Developed MOOC course for the subject "Microcontroller and Embedded Systems Design (KEC-061)" sponsored by AKTU Lucknow, UP at AKTU Digital Education Platform and broadcast at SWAYAM PRABHA for the students. He has more than 28 published papers in Scopus Indexed Journals/ Conferences and 4 other reputed journals along with more than 25 Interna

Shafat Khan is currently serving as an Assistant Professor in the Department of Computer Science at King Khalid University, Saudi Arabia. He holds a Ph.D. in Computer Science with a specialization in Cloud Computing and GIS-based systems, and a Master's in Business Administration. With over 13 years of academic and industry experience, his research interests include cloud service optimization, AI-integrated systems, and IoT-enabled smart technologies. He has published several research papers in reputed journals and has also served as a peer reviewer for international scientific publications.

Mudassir Khan is currently working as Assistant Professor in the department of Computer Science at College of Science & Arts Tanumah, King Khalid University Abha Saudi Arabia.He is pursuing Ph.D. in Computer Science & Engineering from Noida International University Gautam Budh Nagar (NIU) India. He has completed his Graduation and Masters from India.He has more than 7 years of Teaching Experience at the King Khalid University of Saudi Arabia.He has published more than 8 papers in International Journals and conferences. He is the Member of various technical/ professional societies such as IEEE, UACEE, Internet Society,IAENG and CSTA.His research interest includes Big data, IoT, deep learning, Computer Security, Cyber Security and Cloud Computing.

Shabana Mehfuz (Senior Member, IEEE) received the B.Tech. degree in Electrical engineering from Jamia Millia Islamia, India, in 1996, the M.Tech. degree in Computer Technology from IIT Delhi, in 2003, and the Ph.D. degree in Computer Engineering from Jamia Millia Islamia, in 2008. She secured First position in the order of merit in B.Tech.(Electrical Engineering). She has been working at the Department of Electrical Engineering, Jamia Millia Islamia, for the past 24 years. She has published more than 100 articles in International Journals and Conferences. She has guided eight Ph.D.s and is currently supervising other ten candidates. She is a Life Member of ISTE, a member of the Institution of Engineers, and a Life Member of the Computer Society of India. She has received grants for research projects from agencies, such as AICTE and UGC. She has acted as the Track Chair for two flagship International IEEE conferences held in 2015 and 2019. She has been awarded the International Inspirational Women Award 2020 for Best Performer in Government Award by GISR Foundation and. She has been awarded national award for excellence in education 2021 in the category of university/college teachers by AMP organization. Her research interests include Cloud Computing, Internet of Things (IoT), Wireless Sensor Networks, Mobile ad-hoc Network, Artificial Intelligence and Machine Learning.

Firas Tayseer Mohammad Ayasrah is an esteemed scholar and educator specializing in Educational Technology and its applications in teaching and learning. He holds a PhD in Educational Technologies and serves as an Associate Professor at Al Ain University, UAE. With extensive academic experience, he has taught undergraduate and graduate courses in educational technology, instructional design, and digital learning. His research focuses on Artificial Intelligence (AI) in education, digital learning environments, and immersive technologies such as virtual and augmented reality. He explores the integration of AI in instructional systems, its impact on science education, and its applications in higher education. Additionally, his work addresses online learning, educational data analytics, and innovative teaching methodologies. He has published his research findings in reputable international journals and actively participates in research projects exploring AI-driven educational solutions. His contributions aim to enhance technology-enhanced learning, digital pedagogy, and intelligent educational systems, shaping the future of education in the digital age.

Yogeesh N, Government First Grade College, Tumkur, is one of the leading government institutes in Karnataka state among the 412 government colleges across the state. At Government First Grade College, Tumkur, Dr. Yogeesh N, has been working as an assistant professor and head of the department of mathematics since 2013–14. He started his teaching career in the academic year 2005-06 as a lecturer in the College of Diploma in Education, where he taught methodology in mathematics teaching and educational psychology. After that, he worked as a principal and educational officer at the Sacred Heart Group of Institutions from 2007 to 2009. He has now worked in many government-first colleges since 2009, and he has served on the Board of Studies and the Board of Examination at Tumkur University, as well as in many autonomous colleges. He hosted state-level and national-level seminars as organizing secretary, and he also worked in various online and offline journals as a chief editor and editorial member. He wrote several manuals for mathematics labs using FOSS and some academic books, along with paper publications in various journals.

Rohith Bommalla Naresh (IEEE Member) is a Network Lead Engineer with Go8IT, . He specializes in SD-WAN, network automation, and DevOps integration, contributing to advanced network infrastructure solutions. Mr. Naresh is recognized for optimizing network performance across 50+ sites using Cisco Viptela, reducing latency by 25%, 3. He can be contacted at bnr756@gmail.com. His current efforts focus on hybrid cloud architectures and security enhancements. He holds certifications in AWS Certified Solutions Architect – Associate, SailPoint Identity Security Leader, Six Sigma White Belt, Saviynt Identity Security for AI Age, and LinkedIn Learning IT Security Foundations, with CCNA and CKA in progress. He is an active IEEE member, engaging in professional networks.

Zahid Parwez Assistant Professor, Xavier Law School, XIM University, Odisha, India

Index

Symbols

3D-Printed Sensors 2, 12, 15, 16, 27, 28, 39, 40, 41, 42, 43, 44, 45, 47
3D Printing 1, 2, 3, 4, 5, 6, 7, 8, 9, 10, 11, 12, 13, 14, 16, 17, 18, 19, 20, 21, 22, 23, 24, 26, 28, 29, 30, 31, 32, 36, 47, 51, 52, 53, 56, 58, 71, 72, 73, 75, 76, 77, 78, 79, 80, 81, 82, 83, 84, 85, 86, 87, 88, 89, 90, 91, 92, 93

A

Additive Manufacturing 2, 6, 7, 8, 10, 11, 26, 32, 33, 73, 80
Agisoft 151, 152, 153, 154, 155, 161, 166, 168, 169, 171, 172, 173, 174
Agricultural Innovation 17, 71, 76
Agricultural Monitoring 112, 188, 210, 242
Agriculture 5.0 97, 98, 101, 102, 103, 104, 106, 107, 113, 115, 116, 120, 133, 136, 151, 152, 153, 156, 158, 164, 165, 174, 179, 180, 181, 182, 183, 185, 186, 187, 190, 193, 194, 196
AI in Agriculture 110, 111
Algorithmic Bias 54, 57, 61, 118
Artificial Intelligence (AI) 1, 6, 11, 14, 32, 37, 51, 52, 53, 54, 55, 56, 57, 58, 59, 60, 61, 62, 63, 64, 65, 66, 67, 68, 69, 70, 71, 72, 74, 75, 76, 77, 78, 79, 80, 84, 85, 87, 88, 89, 90, 91, 92, 93, 97, 98, 99, 100, 101, 102, 103, 104, 105, 106, 108, 109, 110, 111, 112, 113, 114, 115, 116, 117, 120, 134, 144, 147, 152, 164, 174, 179, 180, 181, 184, 186, 190, 192, 193, 194, 196, 199, 203, 210, 211, 212, 214, 216, 224, 225, 234, 235

C

Cloud Analytics 29
Curvature Accuracy 152
Cybersecurity 94, 148, 179, 180, 181, 182, 183, 184, 185, 186, 187, 188, 190, 191, 192, 193, 195, 196, 197, 198, 199, 200, 201, 225

D

Data Security 32, 38, 75, 136, 197, 238
Decision Support Systems 109, 116, 157, 172, 180
Digital Agriculture 193, 202
Digital Literacy 75, 111
Digital Twins 77, 103, 104, 123, 124, 125, 126, 127, 128, 129, 130, 131, 132, 133, 134, 135, 136, 137, 138, 139, 140, 141, 142, 143, 144, 145, 146, 147, 148, 151, 152, 153, 156, 157, 158, 159, 160, 161, 164, 165, 167, 172, 175, 176, 179, 181, 182, 183, 184, 185, 187, 188, 189, 190, 192, 193, 197, 198, 200, 201, 202, 203, 204, 205, 206, 208, 209, 211, 212, 213, 214, 216, 217, 218, 219, 220, 221, 223, 224, 225, 227, 228, 229, 230, 234, 236, 238, 239, 243, 246, 248, 249
Due Process 51, 52, 53, 54, 55, 58, 60, 61, 62, 64, 65, 66, 67, 69

E

Environmental Monitoring 7
Extended Reality (XR) 70, 71, 72, 73, 74, 76, 84, 88, 89, 94

F

Fair Trial 57, 67
Farmer Empowerment 112
Farm Management 88, 94, 106, 123, 203, 204, 216, 228, 235
Food Processing 1, 2, 6, 7, 8, 11, 24, 32

G

Green Agriculture 73

H

Human-Centric AI 97, 101, 108, 109, 111, 112, 113

I

IoT Sensors 4, 45, 46, 48, 160, 217, 228, 234

J

Judicial Decision-Making 52, 56
Judicial Ethics 51, 52, 53, 54, 55, 61

L

Legal AI 56

O

ODM 151, 152, 166, 167, 168, 169, 170, 171, 172, 174

P

pH 3, 28, 30, 31, 36, 39, 42, 43, 45, 46, 48, 125, 134, 212, 234, 237
Photogrammetry 152, 154, 156, 157, 161, 162, 163, 164, 165, 166
Pix4D 153, 154, 155, 161, 164, 169, 170
Precision Agriculture 1, 2, 3, 5, 8, 9, 16, 18, 22, 26, 31, 32, 39, 41, 44, 76, 78, 79, 83, 104, 110, 121, 122, 125, 129, 146, 160, 164, 183, 195, 202, 203, 204, 206, 208, 211, 214, 216, 219, 223, 225
Precision Farming 7, 11, 28, 31, 72, 83, 88, 101, 123, 128, 135, 152, 169, 170, 203, 205, 209, 210, 211, 216, 228, 230, 231, 232, 237

R

Real-Time Monitoring 46, 123, 142, 187, 188, 198, 243
Resource Optimization 71, 89, 92, 164, 189

S

Smart Agriculture 1, 32, 35, 107, 121, 123, 127, 129, 130, 133, 135, 136, 138, 139, 141, 142, 177, 191, 193, 200, 225, 228, 230, 231, 232, 233, 241, 242, 244, 245, 246, 247, 248, 249
Smart Farming 9, 12, 27, 32, 33, 39, 41, 42, 45, 47, 75, 97, 98, 100, 101, 102, 106, 116, 118, 121, 126, 128, 130, 133, 135, 139, 141, 176, 179, 180, 184, 187, 194, 197, 200, 208, 209, 225, 228, 231, 233, 244, 245, 246, 248, 249
Soil Moisture 2, 11, 30, 36, 38, 42, 43, 45, 46, 99, 133, 135, 139, 157, 160, 189, 198, 230, 232, 234, 241, 242, 243
Spatial Modeling 151
Sustainable Agriculture 72, 79, 98, 106, 124, 225, 227, 228, 234, 236, 239, 243
Sustainable Farming 71, 76, 82, 89, 91, 170, 214, 249

T

Threat Detection 194, 195

V

Virtual Environment 125, 212, 234, 238

Y

Yield Improvement 41, 71, 141
Yield Optimization 183